Everyday Mathematics®

The University of Chicago School Mathematics Project

STUDENT REFERENCE BOOK

McGraw Hill Education

The University of Chicago School Mathematics Project

Max Bell, Director, *Everyday Mathematics* First Edition
James McBride, Director, *Everyday Mathematics* Second Edition
Andy Isaacs, Director, *Everyday Mathematics* Third, CCSS, and Fourth Editions
Amy Dillard, Associate Director, *Everyday Mathematics* Third Edition
Rachel Malpass McCall, Associate Director, *Everyday Mathematics* CCSS and Fourth Editions
Mary Ellen Dairyko, Associate Director, *Everyday Mathematics* Fourth Edition

Authors
Max Bell
Jean Bell
John Bretzlauf
Amy Dillard
James Flanders
Robert Hartfield
Andy Isaacs
Catherine Randall Kelso
James McBride
Kathleen Pitvorec
Peter Saecker

Writers
Lisa J. Bernstein
Jeanne Di Domenico
Andy Carter
Gina Garza-Kling
Lila K.S. Goldstein
Jesch Reyes
Elizabet Spaepen
Judith S. Zawojewski

Digital Development Team
Carla Agard-Strickland, Leader
John Benson
Gregory Berns-Leone
Scott Steketee

Technical Art
Diana Barrie, Senior Artist
Cherry Inthalangsy

UCSMP Editorial
Elizabeth Olin
Kristen Pasmore
Molly Potnick

Contributors
Lance Campbell
Mary Ellen Dairyko
Rosalie A. DeFino
Kathryn Flores

www.everydaymath.com

Send all inquiries to:
McGraw-Hill Education
8787 Orion Place
Columbus, OH 43240

ISBN: 978-0-02-138355-9
MHID: 0-02-138355-3

Printed in the United States of America.

4 5 6 7 8 9 DOW 20 19 18 17 16

Contents

Standards for Mathematical Practice   1

Operations and Algebraic Thinking   37

Contents

Number and Operations in Base Ten   85

Number and Operations—Fractions   131

Measurement and Data   163

Contents

Contents

Real-World Data   263

Contents

About the *Student Reference Book*

A reference book is organized to help readers find information quickly and easily. Dictionaries, encyclopedias, atlases, and cookbooks are examples of reference books. You can look in reference books to find the information you need.

The *Everyday Mathematics Student Reference Book* is a book that you can use everyday. You can use the *Student Reference Book* to learn about new topics in math, to review ideas and skills you learned before, and to find the meanings of math words.

This *Student Reference Book* has the following information:

- A **table of contents** listing the topics covered and giving an overview of how the book is organized

- Essays describing **mathematical practices**

- Essays on **mathematical content**

- A collection of tables, charts, diagrams, posters, and maps that include **real-world data**

- A collection of **photo essays** that show in words and pictures some of the ways that mathematics is used

- Directions on how to play **mathematical games** to practice your math skills

- An **appendix** that includes directions on how to use a **calculator** for some mathematical operations

- A **glossary** of mathematical terms with definitions and some illustrations

- An **answer key** for **Check Your Understanding** problems

- An **index** to help you find topics quickly

- **Videos** and **interactive problems** available through the electronic version of this book in the Student Learning Center

How to Use the *Student Reference Book*

As you work in class or at home, you can use the *Student Reference Book* to help you solve problems. For example, when you don't remember the meaning of a word or aren't sure what method to use, you can use the *Student Reference Book* as a tool.

You can look in the **table of contents** or **index** to find pages that may help you solve the problem. The pages may have definitions of math words and show examples of problems that are similar to yours.

At the end of many of the essays, you will find problems in **Check Your Understanding** boxes. These exercises can help you know whether you understand the mathematics on the pages. After you solve these problems, you can often check your answers with the **answer key** at the back of the book.

The **Standards for Mathematical Practice** section includes interesting problems that you can solve. The discussions show how third-grade students use the practices to solve these problems.

The world of mathematics is a very interesting and exciting place. Read the **photo essays,** explore **real-world data,** or review topics learned in class. The *Student Reference Book* is a great place to continue your investigation of math topics and ideas.

The more you explore this book, the more you will know where to find information that can help you better understand mathematics.

Standards for Mathematical Practice

Mathematical Practices

Mathematical practices are ways that people think and work when they use mathematics to solve problems in their jobs and at home.

In what ways are these children thinking and working to solve problems?

In this section, you will see how some children use mathematical practices as they solve problems. As you read, think about ways the practices can help you become a better problem solver.

Here are the Standards for Mathematical Practice you will use in *Everyday Mathematics*. Below each standard are Goals for Mathematical Practice (GMPs) that can help you understand what it means to use each of the practices.

Mathematical Practice 1: Make sense of problems and persevere in solving them.

> GMP1.1 Make sense of your problem.
>
> GMP1.2 Reflect on your thinking as you solve your problem.
>
> GMP1.3 Keep trying when your problem is hard.
>
> GMP1.4 Check whether your answer makes sense.
>
> GMP1.5 Solve problems in more than one way.
>
> GMP1.6 Compare the strategies you and others use.

Mathematical Practice 2: Reason abstractly and quantitatively.

GMP2.1 Create mathematical representations using numbers, words, pictures, symbols, gestures, tables, graphs, and concrete objects.

GMP2.2 Make sense of the representations you and others use.

GMP2.3 Make connections between representations.

Mathematical Practice 3: Construct viable arguments and critique the reasoning of others.

GMP3.1 Make mathematical conjectures and arguments.

GMP3.2 Make sense of others' mathematical thinking.

Mathematical Practice 4: Model with mathematics.

GMP4.1 Model real-world situations using graphs, drawings, tables, symbols, numbers, diagrams, and other representations.

GMP4.2 Use mathematical models to solve problems and answer questions.

Mathematical Practice 5: Use appropriate tools strategically.

GMP5.1 Choose appropriate tools.

GMP5.2 Use tools effectively and make sense of your results.

Mathematical Practice 6: Attend to precision.

GMP6.1 Explain your mathematical thinking clearly and precisely.

GMP6.2 Use an appropriate level of precision for your problem.

GMP6.3 Use clear labels, units, and mathematical language.

GMP6.4 Think about accuracy and efficiency when you count, measure, and calculate.

Mathematical Practice 7: Look for and make use of structure.

GMP7.1 Look for mathematical structures such as categories, patterns, and properties.

GMP7.2 Use structures to solve problems and answer questions.

Mathematical Practice 8: Look for and express regularity in repeated reasoning.

GMP8.1 Create and justify rules, shortcuts, and generalizations.

Problem Solving: Make Sense and Keep Trying

Third graders are solving this problem:

There are 11 animals and 38 legs in the barnyard. The animals are chickens and cows. How many of each animal is in the barnyard?

Noah draws 11 circles, one for each animal. He puts 4 lines for legs inside 2 circles to show cows and 2 lines for legs inside 2 circles to show chickens.

How would you solve this problem?

So far I have 12 legs. How am I going to get to 38 legs?

Noah

GMP1.1 Make sense of your problem.

Noah made sense of this problem when he drew 11 circles for animals and then drew 2 legs for chickens and 4 legs for cows. He knew he needed to get to 38 legs.

Abby starts with a good guess. She first guesses 6 cows and 5 chickens because there are 11 animals in all. Then she writes:

6 cows + 5 chickens

6 × 4 legs + 5 × 2 legs
24 legs + 10 legs = 34 legs

That makes just 34 legs, which isn't enough legs. We need more cows. So let's trade a chicken for a cow.

Abby

GMP1.2 Reflect on your thinking as you solve your problem.

Abby reflected on her thinking when she saw that she didn't have enough legs. She knew she needed more cows.

Abby tries another guess to get closer to 38 legs. She tries one more cow and one less chicken.

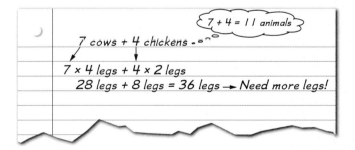

7 + 4 = 11 animals

7 cows + 4 chickens

7 × 4 legs + 4 × 2 legs
28 legs + 8 legs = 36 legs → Need more legs!

Abby thinks about her new guess. She realizes that there are still not enough legs. So she tries again with 8 cows and 3 chickens.

8 cows + 3 chickens

8 + 3 = 11 animals

8 × 4 legs + 3 × 2 legs
32 legs + 6 legs = 38 legs

This is a hard problem! But I figured it out. My last guess of 8 cows and 3 chickens works. I have 11 animals and 38 legs.

Abby

GMP1.3 Keep trying when your problem is hard.

Instead of giving up, Abby kept trying even when she didn't find the answer right away. Each time she made a guess and checked it, she thought about how to make a better guess the next time.

Noah and Abby check Abby's answer by adding to Noah's picture.

cows

4   8   12   16   20   24   ← 8 cows

28   32   34   36   38   ← 3 chickens

chickens

Now my picture works. When I count all the legs for 8 cows and 3 chickens, there are 38 legs in all.

Noah

GMP1.4 Check whether your answer makes sense.

Noah's drawing showed that Abby's answer worked, so they knew the answer made sense.

GMP1.5 Solve problems in more than one way.

Abby made good guesses and used number models. Noah and Abby drew a picture that fit the problem.

Tiffany solves the problem a different way. She makes a table that shows the numbers of cows and chickens that add up to 11.

Cows	Chickens	Legs	Total Legs
1	10	$1 \times 4 + 10 \times 2$	24
3	8	$3 \times 4 + 8 \times 2$	28
5	6	$5 \times 4 + 6 \times 2$	32
7	4	$7 \times 4 + 4 \times 2$	36
9	2	$9 \times 4 + 2 \times 2$	40
8	3	$8 \times 4 + 3 \times 2$	38

My table is like Abby's number models, but the table helps me keep track of my guesses. And we get the same answer: 8 cows and 3 chickens have 38 legs in all.

Tiffany

GMP1.6 Compare the strategies you and others use.

Tiffany compared her strategy to Abby's. She noticed that they both used number sentences and got the same answer.

Mathematical Practice 1: Make sense of problems and persevere in solving them.

Create and Make Sense of Representations

Third graders are solving the problem:

$$604 - 381 = ?$$

Look carefully at the mathematical **representations** the children use to answer the question and share their thinking.

How would you solve this problem? How would you show your thinking?

Tyler knows that he can solve a subtraction problem by counting up. He thinks of the problem in two ways:

$$604 - 381 = ? \quad \text{and} \quad 381 + ? = 604$$

He draws an open number line and starts at 381.

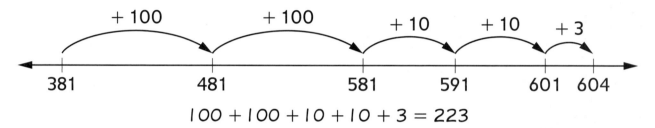

$$100 + 100 + 10 + 10 + 3 = 223$$

I use numbers that are easy to add. I count up 100 to get to 481, and then 100 more to get to 581. It is easy to add 10 to get to 591 and 10 more to get to 601. I need 3 more to get to 604. Then I add my jumps and get 223.

GMP2.1 Create mathematical representations using numbers, words, pictures, symbols, gestures, tables, graphs, and concrete objects.

Tyler

GMP2.2 Make sense of the representations you and others use.

Tyler created a mathematical representation when he drew an open number line. He made sense of his representation by thinking about how to use friendly numbers to count up on the open number line to solve the problem.

Bella uses numbers and symbols to show expand-and-trade subtraction, and Abby uses base-10 blocks. They show their steps to the class.

I can write the problem using expanded form so I see the hundreds, tens, and ones.

I show 6 hundreds and 4 ones using 6 flats and 4 cubes.

Bella

Abby

$$604 \rightarrow 600 + 0 + 4$$
$$-381 \rightarrow 300 + 80 + 1$$

I notice that Bella writes a 0 in the tens place to show 0 tens in 604, and Abby shows 0 tens by not using any longs.

Martin

Bella and Abby decide they need to make a trade. They each show their trade to the class.

600 is more than 300, and 4 is more than 1. But 0 is less than 80, so I need to trade. I trade a 100 from 600 for ten 10s or 100. So I write 500 above the 600 and 100 in the tens place.

I can take away 3 flats for 300 and 1 cube for 1. But I don't have any longs to subtract 80. So first I trade 1 flat for 10 longs.

Bella

Abby

$$604 \rightarrow \cancel{600} + \cancel{0} + 4$$
$$-381 \rightarrow 300 + 80 + 1$$

(above: 500   100)

Bella trades 1 hundred from the hundreds column for 10 tens, or 100, in the tens column, and Abby trades 1 flat for 10 longs. They both trade 1 hundred for 10 tens.

Martin

Abby and Bella show how they subtract to find their final answers.

> After subtracting, I have 200 + 20 + 3 = 223, so that's the answer.

> Now I can take away 8 longs. So I'm left with 2 hundreds, 2 tens, and 3 ones, and that's 223. So 604 − 381 = 223.

Bella

Abby

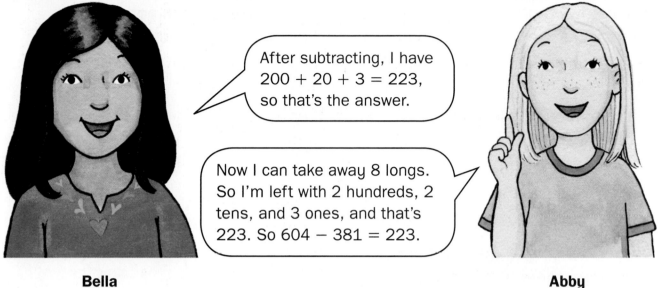

$$604 \rightarrow \overset{500}{\cancel{600}} + \overset{100}{\cancel{0}} + 4$$
$$- 381 \rightarrow 300 + 80 + 1$$
$$ 200 + 20 + 3 = 223$$

> Bella and Abby use different representations. One uses numbers and symbols and the other uses base-10 blocks. They both started with 604 and made trades to subtract. Their answers are the same.

Martin

GMP2.3 Make connections between representations.

Tyler, Bella, and Abby created different mathematical representations to solve this problem. Martin made sense of Bella's and Abby's representations and described connections between them by thinking about how they both used place value.

Mathematical Practice 2: Reason abstractly and quantitatively.

Make Conjectures and Arguments

When solving problems, people often start by making a **conjecture**. A **conjecture** is a statement that might be true. In mathematics, conjectures are not simply guesses; they are based on some information or mathematical thinking.

Mathematical reasoning that shows whether a conjecture is true or false is called an **argument**. You can use words, pictures, and symbols when you make a mathematical argument.

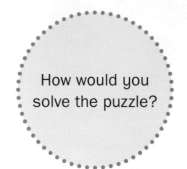

How would you solve the puzzle?

Martin and Tiffany are trying to solve this puzzle:

Clue 1: When you make groups of 5 erasers, there is 1 left over.

Clue 2: When you make groups of 3 erasers, there are 2 left over.

Find a number of erasers that fits these clues.

I think 6 erasers works because I can make one group of 5 erasers with 1 left over.

Martin

Martin's conjecture: An answer to the puzzle is 6 erasers.

Martin's argument: The conjecture is true because it fits Clue 1. When you make a group of 5 erasers, there is 1 left over.

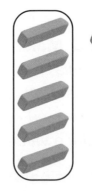

I disagree because of Clue 2. If I make groups of 3 with 6 erasers, there will be 0 left over. Clue 2 says there should be 2 left over.

Tiffany

Martin realizes he was wrong. He uses erasers to find a number that fits both clues.

11 erasers fits Clue 1.   11 erasers fits Clue 2.

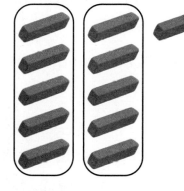   

Now I think 11 erasers works. That fits both clues.

When I make groups of 5 erasers, there is 1 left over, and when I make groups of 3 erasers, there are 2 left over.

Martin

Martin's new conjecture: An answer to the puzzle is 11 erasers.

Martin's new argument: The conjecture is true because it fits *both* clues. When 11 erasers are grouped by 5, there is 1 left over. When 11 erasers are grouped by 3, there are 2 left over.

GMP3.1 Make mathematical conjectures and arguments.
Martin made a conjecture that he thought was true, but his argument didn't show that 6 erasers worked for both clues. Tiffany argued that his conjecture was false by showing that 6 erasers doesn't fit Clue 2.

For a conjecture to be true, it must always *be true. To show that a conjecture is false, all you need is one example where the conjecture doesn't work.*

GMP3.2 Make sense of others' mathematical thinking.
Tiffany made sense of Martin's mathematical thinking when she tested his conjecture. When Martin made sense of Tiffany's thinking, he revised his conjecture so that it would be true.

Mathematical Practice 3: Construct viable arguments and critique the reasoning of others.

Check Your Understanding

Use what you learned about making conjectures and arguments to answer these questions:

1. What did Tiffany need to do to show that Martin's conjecture was false?

2. How did Martin show that his new conjecture was true?

3. Make another conjecture and argument for the problem. Find another number that fits both clues.

Check your answers in the Answer Key.

Create and Use Mathematical Models

Third graders are solving this problem:

Ken planted red and yellow flowers in the school garden.

For every 3 red flowers Ken planted, he planted 2 yellow flowers.

If Ken planted 25 flowers all together, how many of each color did he plant?

The third graders use mathematical **models** to help them solve the problem. They draw pictures and write number models that fit the problem and show their thinking.

How would you solve the problem? How would you show your thinking?

Tiffany uses a drawing to model the flowers.

> I can draw groups of 3 red and 2 yellow flowers over and over again until I get to 25 total flowers.

Tiffany

Red:

Yellow:

Alex thinks that the number of red flowers is one more than the number of yellow flowers. He uses letters to stand for 13 red and 12 yellow flowers.

R R R R R R R R R R R R R
Y Y Y Y Y Y Y Y Y Y Y Y

13 red flowers + 12 yellow flowers
= 25 flowers planted

> Does my model make sense? I'm not sure it fits the problem.

Alex

Tiffany adds to her model in a green marker as she explains her thinking.

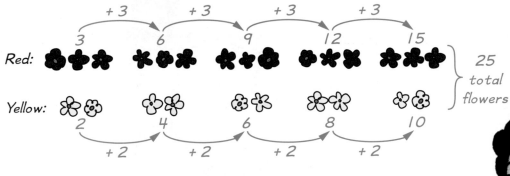

Red:
+3 +3 +3 +3
3   6   9   12   15

25 total flowers

Yellow:
2   4   6   8   10
+2 +2 +2 +2

> I see that each group of 3 red and 2 yellow flowers is 5 flowers. So I count by 5s until I get to 25 flowers total. Then I count by 3s to find that there are 15 reds and count by 2s to find that there are 10 yellows.

Tiffany

Alex thinks about Tiffany's drawing and sees that his drawing doesn't fit the problem. He explains, "I didn't think about how the red and yellow flowers were grouped. Tiffany's drawing groups 2 yellow with 3 red flowers. It shows that 15 red and 10 yellow flowers were planted. That makes sense because 15 + 10 = 25, and the problem said Ken planted 25 flowers."

GMP4.1 Model real-world situations using graphs, drawings, tables, symbols, numbers, diagrams, and other representations.

Alex and Tiffany both modeled the real-world situation by drawing pictures and writing number models. Alex realized that his model didn't fit the problem because he didn't show the way the flowers were grouped.

GMP4.2 Use mathematical models to solve problems and answer questions.

As Tiffany explained her thinking to Alex, she improved her model by adding arrows to show more clearly how many red and yellow flowers were planted.

Mathematical Practice 4: Model with mathematics.

Check Your Understanding

Martin models the real-world situation using a table. He notices that he gets the same answer as Tiffany:

Red	Yellow	Total
3	2	5
6	4	10
9	6	15
12	8	20
15	10	25

(Red column: +3 between each; Yellow column: +2 between each.)

Make sense of Martin's model.

1. Choose a row. Explain what each number represents in that row.

2. In what ways are Martin's and Tiffany's models the same? In what ways are they different?

Check your answers in the Answer Key.

Choose Tools to Solve Problems

You can use different types of tools to solve problems in mathematics.

Mr. Lopez asks his third-grade class to solve this problem:

Yesterday, at 4:48 P.M., I put enough money in a parking meter for 1 hour and 30 minutes. I returned to my car at 6:09 P.M.

Did I get back to my car before the time on the parking meter ran out? Show how you know.

Noah chooses a toolkit clock to help him solve the problem.

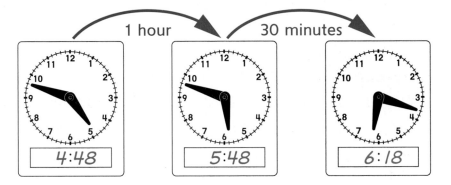

I first show 4:48 on the clock. I add 1 hour to get to 5:48. Then I add 30 minutes, which is halfway around the clock, to get to 6:18.

6:18 is when the parking meter time ended. That's later than 6:09, so Mr. Lopez made it in time.

Noah

Bella uses an open number line with familiar times and a number model.

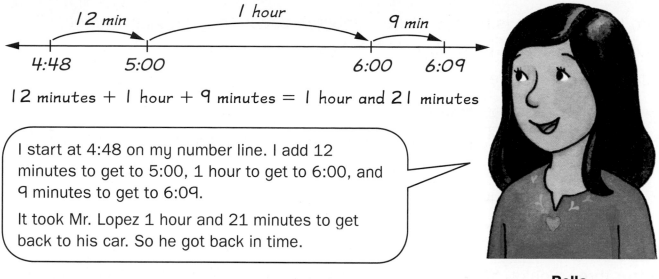

12 minutes + 1 hour + 9 minutes = 1 hour and 21 minutes

I start at 4:48 on my number line. I add 12 minutes to get to 5:00, 1 hour to get to 6:00, and 9 minutes to get to 6:09.

It took Mr. Lopez 1 hour and 21 minutes to get back to his car. So he got back in time.

Bella

GMP5.1 Choose appropriate tools.

Noah and Bella chose tools that helped them solve the problem. Noah used a toolkit clock, and Bella used an open number line and a number model.

Tyler chooses to use a table to solve the problem.

Times	Hours	Minutes
4:48–5:48	1 hour	
5:48–6:00		12 minutes
6:00–6:09		9 minutes
Total	1 hour	21 minutes

First, I go from 4:48 to 5:48. That makes 1 hour. I add 12 minutes to get to 6:00 and 9 more minutes to get to 6:09.

Although Bella and I use different tools, the times we add and the total time are the same. We agree that Mr. Lopez got back in time.

Tyler

GMP5.2 Use tools effectively and make sense of your results.

Tyler used a table to solve the problem. The table helped him make sense of Bella's number line and see how he found the same answer as Bella.

Mathematical Practice 5: Use appropriate tools strategically.

Be Precise and Accurate

Mr. Lopez's class is recording the number of pages they read each week. Each child's goal is to read at least 500 pages in four weeks.

Martin and Bella are keeping track of the number of pages they read.

Martin

Week	Number of Pages
1	147
2	203
3	198
4	

Bella

Week	Number of Pages
1	157
2	149
3	191
4	

After three weeks, Mr. Lopez asks Martin and Bella, "Have you each met the goal of 500 pages yet? If not, how many more pages do you need to read?"

How would you figure out if Martin and Bella have each read 500 pages after three weeks?

I know I read at least 500. I add 1 + 2 + 1 and that's 4. Then I look at the other places.

Martin

Mr. Lopez asks Martin to explain his thinking more clearly using place-value language and units.

Martin thinks about what he wants to say and explains his thinking this way: "First, I add the hundreds digits of my numbers: 1 + 2 + 1 = 4, so that means I read at least 400 pages. Just by looking at the numbers in the tens and the ones places, I know they add up to more than 100 pages. So I've already read more than 500 pages."

GMP6.1 Explain your mathematical thinking clearly and precisely.
Martin explained what he did more clearly when he talked about hundreds, tens, and ones.

GMP6.3 Use clear labels, units, and mathematical language.
Martin's second explanation was clearer because he used place-value language and pages as units.

Bella uses close-but-easier numbers to estimate the number of pages she read.

Bella

Week	Number of Pages
1	157
2	149
3	191
4	

157 → 150
149 → 150
191 → 200

$150 + 150 + 200 = 500$

Bella explains her thinking this way:

"150 + 150 = 300 That's easy to do in my head.
300 + 200 is 500 So I've read about 500 pages.
But I don't know if I'm over or under 500 pages."

Bella uses a calculator to find an exact answer. She adds 157 + 149 + 191:

I got 397. That doesn't make sense because it's almost 100 pages away from my estimate. I must have made a mistake on my calculator. I'd better try it again.

Bella

This time I got 497. That's really close to my estimate of 500 pages, so I think it's right. I only need to read 3 more pages to meet my goal.

Bella

GMP6.2 Use an appropriate level of precision for your problem.

When Bella realized that her estimate didn't tell her if she met her goal, she had to find a more exact answer.

GMP6.4 Think about accuracy and efficiency when you count, measure, and calculate.

Bella found that a calculator was efficient, but she had to think about whether the answer she found with the calculator was accurate.

Mathematical Practice 6: Attend to precision.

Check Your Understanding

Noah keeps track of the pages he reads each week.

Week	Number of Pages
1	97
2	211
3	143
4	

Did Noah read 500 pages in three weeks? Use clear mathematical language and units to explain your thinking.

Check your answers in the Answer Key.

Look for Structure in Mathematics

Third graders are discussing how they can figure out the number of dots on this Quick Look card without counting each one.

Abby sees a pattern.

+4 +4 +4
4, 8, 12, 16

How many dots do you see? How can you tell the number of dots without counting one by one?

The dots are in groups of 4. So I can skip count, adding 4 each time: 4, 8, 12, 16. There are 16 dots in all.

Abby

Bella sees 4 groups of 4. She uses a multiplication fact.

4 groups of 4
$4 \times 4 = 16$

Bella

Noah and Bella look closely at each ten frame.

8   8

It's easy to see that each ten frame shows 8 dots. Two groups of eight is $2 \times 8 = 16$.

Noah

2   2

$20 - 4 = 16$

Here's another way to see 16: 2 full ten frames is 20 in all. I see 2 empty squares in each, so there are 4 squares without dots. So, $20 - 4 = 16$.

Bella

Alex and Tyler see that the dots form rows and columns in an array.

Alex

I look across the 2 ten frames and see 2 arrays. Each has 2 rows and 4 dots in each row. That makes 8 dots in each array, and 8 + 8 = 16.

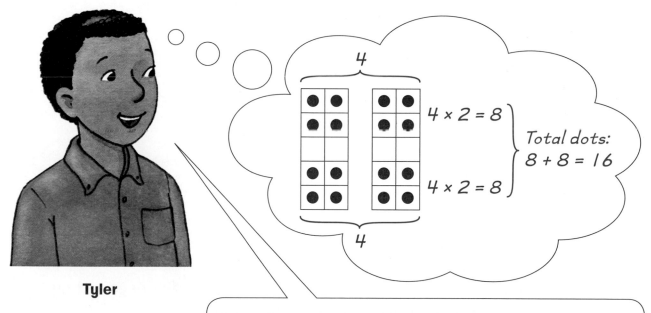

Tyler

I see the same arrays, but I think of each array as 4 columns with 2 dots in each column. I multiply 4 × 2 instead of 2 × 4, and we both get 8 dots in each array. That's the turn-around rule.

GMP7.1 Look for mathematical structures such as categories, patterns, and properties.

Abby found a pattern when she saw you could add 4 four times. Noah saw 2 groups of 8. Bella saw 4 equal groups of 4 and she also saw 20 minus 4. Alex saw 2 arrays, each with 2 rows of 4 dots. Tyler saw that because the dots were in an array, "2 rows of 4" could be thought of as "4 columns of 2."

GMP7.2 Use structures to solve problems and answer questions.

Abby used skip counting to find the total based on the pattern of repeated addition. Noah used multiplication to find the total. Bella used multiplication and subtraction. Alex and Tyler saw arrays. Tyler used the turn-around rule for multiplication to find the total dots in each array.

All of the children used structures to figure out the total number of dots without counting one by one.

Mathematical Practice 7: Look for and make use of structure.

Check Your Understanding

1. How can you figure out the number of dots on this Quick Look card without counting them one by one? Tell how many dots you see and how you see them.

2. Think about your strategy. Is it like one of the strategies described by one of the third graders? If so, which one? Explain.

Check your answers in the Answer Key.

Create and Justify Shortcuts

You can solve math problems easier and faster when you create and justify strategies that can work as shortcuts for many problems.

Tiffany knows her multiplication facts for 5. She says she uses multiplication facts for 5 to help her figure out multiplication facts for 6. Mr. Lopez asks Tiffany to show and explain her strategy to the class.

$6 \times 3 = ?$

$5 \times 3 + 1 \times 3 = ?$
$15 + 3 = 18$

So, $6 \times 3 = 18$.

6 groups of 3 is the same as 5 groups of 3 plus 1 group of 3.

Since 5 groups of 3 is 15, I can add 1 more group of 3 to get 6 groups of 3.

Tiffany

$6 \times 4 = ?$

$5 \times 4 + 1 \times 4 = ?$
$20 + 4 = 24$

So, $6 \times 4 = 24$.

6 groups of 4 is the same as 5 groups of 4 plus 1 group of 4.

Tiffany

Mr. Lopez asks his students to use Tiffany's shortcut strategy on one of the following problems:

$$6 \times 8 = ? \quad 6 \times 7 = ? \quad 6 \times 9 = ?$$

Alex explains and shows how he figured out the first problem: $6 \times 8 = ?$

6 × 8 = ?

5 × 8 = 40 and
1 × 8 = 8, so
6 × 8 = 40 + 8
6 × 8 = 48

6 × 8 means 6 groups of 8.

So I can think of
5 groups of 8 and
1 group of 8.

Alex

Mr. Lopez asks Alex to **justify** or show how and why this shortcut strategy works.

Alex draws 6 rows of 8 Xs to show 6×8. Then he labels his drawing.

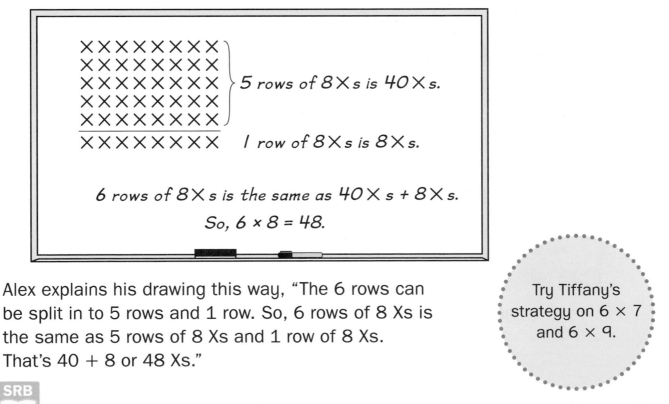

X X X X X X X X
X X X X X X X X
X X X X X X X X ⎫ 5 rows of 8 Xs is 40 Xs.
X X X X X X X X
X X X X X X X X ⎭
X X X X X X X X   1 row of 8 Xs is 8 Xs.

6 rows of 8 Xs is the same as 40 Xs + 8 Xs.
So, 6 × 8 = 48.

Alex explains his drawing this way, "The 6 rows can be split in to 5 rows and 1 row. So, 6 rows of 8 Xs is the same as 5 rows of 8 Xs and 1 row of 8 Xs. That's 40 + 8 or 48 Xs."

Try Tiffany's strategy on 6 × 7 and 6 × 9.

Abby notices that this shortcut strategy works for other multiplication facts. She states her shortcut this way:

"A shortcut strategy for finding the answer to any multiplication fact is to use a helper fact. You can break the groups into smaller groups and then add their products."

She gives an example of her shortcut as she writes on the board.

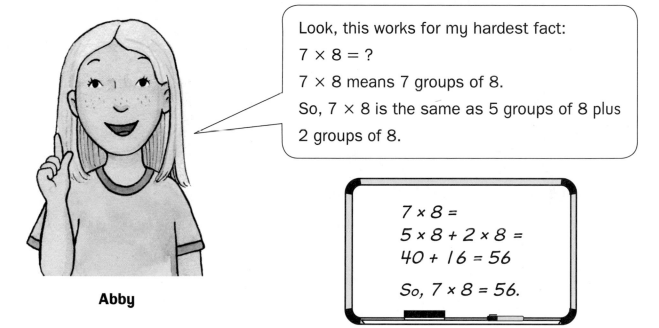

Look, this works for my hardest fact:

$7 \times 8 = ?$

7×8 means 7 groups of 8.

So, 7×8 is the same as 5 groups of 8 plus 2 groups of 8.

Abby

$7 \times 8 =$
$5 \times 8 + 2 \times 8 =$
$40 + 16 = 56$

So, $7 \times 8 = 56$.

GMP8.1 Create and justify rules, shortcuts, and generalizations.
Tiffany created a shortcut strategy when she figured out a way to find answers to multiplication facts with 6 using what she knows about multiplication facts with 5.

Alex justified the shortcut when he showed how and why the strategy works for 6×8.

Abby made a **generalization** *when she described how this shortcut works for other multiplication facts.*

Mathematical Practice 8: Look for and express regularity in repeated reasoning.

Guide to Solving Number Stories

Solving number stories is a big part of mathematics. Good problem solvers often follow a few simple steps every time they solve a number story. When you are solving a number story, you can follow these steps in any order.

Make sense of the problem.

• Read the problem. What do you understand?

• What do you know?

• What do you need to find out?

Make a plan.

What will you do?

• Add?   • Make a table?

• Subtract?   • Make a graph?

• Multiply?   • Write a number model?

• Divide?   • Use counters or base-10 blocks?

• Draw a picture?   • Make tallies?

• Draw a diagram?   • Use a number grid or number line?

Solve the problem.

• Show your work.

• Keep trying if you get stuck.

→ Reread the problem.

→ Think about the strategies you have already tried and try new strategies.

• Write your answer with the units.

Check.

Does your answer make sense? How do you know?

Check Your Understanding

Solve the number stories.

1. Jennifer had 40 pennies. She put 25 pennies in her bank and gave the other pennies to 3 friends. How many pennies did each friend get if they got equal shares?

2. Mr. Cohen's third-grade class has more boys than girls. If there were 2 more boys, there would be twice as many boys as girls. There are 8 girls in the class. How many boys are there?

3. The product of 2 numbers is 12. The sum of the numbers is 7. What are the numbers?

4. Rashid bought 4 markers for 56¢. How much did each marker cost?

Check your answers in the Answer Key.

A Problem-Solving Diagram

The diagram below can help you think about problem solving. The boxes show the kinds of things you do when you use mathematical practices to solve problems. The arrows show that you don't always do things in the same order.

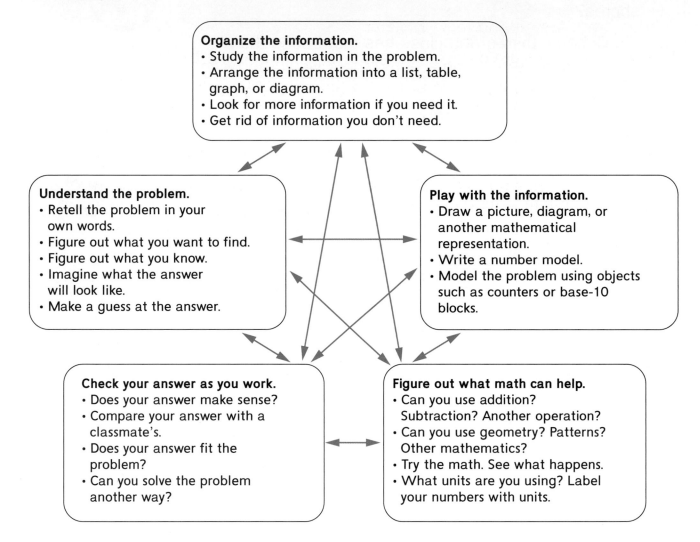

Organize the information.
- Study the information in the problem.
- Arrange the information into a list, table, graph, or diagram.
- Look for more information if you need it.
- Get rid of information you don't need.

Understand the problem.
- Retell the problem in your own words.
- Figure out what you want to find.
- Figure out what you know.
- Imagine what the answer will look like.
- Make a guess at the answer.

Play with the information.
- Draw a picture, diagram, or another mathematical representation.
- Write a number model.
- Model the problem using objects such as counters or base-10 blocks.

Check your answer as you work.
- Does your answer make sense?
- Compare your answer with a classmate's.
- Does your answer fit the problem?
- Can you solve the problem another way?

Figure out what math can help.
- Can you use addition? Subtraction? Another operation?
- Can you use geometry? Patterns? Other mathematics?
- Try the math. See what happens.
- What units are you using? Label your numbers with units.

Problem Solving and the Mathematical Practices

When solving problems, you can use mathematical practices to help make sense of the problem and to keep trying when you are not sure what to do next.

Third graders are solving this problem:

A frog is at the bottom of a box that is 5 feet deep. At the beginning of each hour, the frog jumps up 3 feet onto the side of the box. The frog then takes a break for the rest of the hour. During the break, the frog slips down 2 feet. This happens over and over until the frog reaches the top edge of the box, where he can climb out and leap down to the ground.

When will the frog reach the top edge of the box so that he can climb out?

How would you start to solve this problem?

After children have worked on the problem, they show and explain their solutions to the class.

The frog moves 1 foot each hour. I know that because he jumps up 3 feet and slips back 2 each time and 3 − 2 = 1 foot. So, 5 feet will take 5 hours.

Tiffany and I worked together, and we think that the frog gets out in 4 hours. I'll show you how we figured it out.

Tyler

Alex

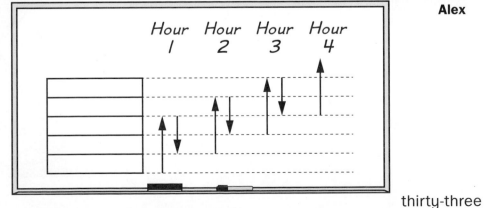

I don't understand. What do the dotted lines and arrows mean?

To help the children in the class make sense of Alex and Tiffany's representation, Mr. Lopez asks them to use labels so that their explanation is clearer and more precise. Tiffany adds to the diagram using a red marker.

Tyler

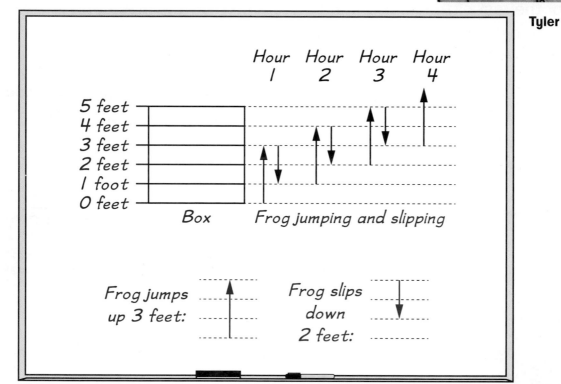

Now I get it. But I think your diagram shows the frog getting out at the beginning of the third hour. See? The frog is at the top of the box at the beginning of Hour 3.

Alex, Tiffany, and other children in the class are not sure they agree with Tyler. So Mr. Lopez asks the children to test Tyler's conjecture and argument.

Tyler

Tiffany and Alex choose a different tool, a chart, to solve the problem a different way. Mr. Lopez asks Tiffany and Alex to show and explain their chart to the class.

☾ Hour	Jumps up 3 feet at the beginning of an hour Where is the frog?	Slips 2 feet down by the end of the hour Where is the frog?
1st hour	3 feet	3 feet − 2 feet = 1 foot
2nd hour	1 foot + 3 feet = 4 feet	4 feet − 2 feet = 2 feet
3rd hour	2 feet + 3 feet = 5 feet	
4th hour		

Tyler is right. The chart shows the exact time when the frog jumped up to 5 feet.

At the beginning of the third hour, the frog makes it to the top edge of the box, so he can climb out. He doesn't even slip back.

Tiffany

Now I see what Tyler was saying about our first diagram. It does show that the frog made it to 5 feet at the beginning of Hour 3, and that is the same as the row for the third hour in the chart. Using our diagram, I didn't think it looked like the frog was really *out* yet. So using the chart helps me make sense of our diagram.

Alex

Check Your Understanding

Use the mathematical practices listed on pages 2 and 3 to answer the questions below. It may help to look at the examples on pages 4–29.

1. What mathematical practices did Tiffany and Alex use to solve the problem about the frog jumping out of the box?

2. What mathematical practices did Tyler use?

Tiffany   Alex   Tyler

Operations and Algebraic Thinking

Mary-Ella Keith/Alamy

Multiplication and Equal Groups

Some number stories are about **equal groups.** Groups are **equal** if they all contain the same number of items. You can use multiplication to find the total number of items in a set of equal groups.

When two numbers are multiplied, the number answer is called the **product.** The two numbers that are multiplied are called **factors** of the product.

Example

Find the total number of dots shown.

Think of equal groups of dots. There are 3 *groups* and 4 dots *in each group.* So this is an equal-groups problem.

One way to find the total number of dots is to count by ones. There are 12 dots. Another way to find the total number of dots is to add the dots from each die: $4 + 4 + 4 = 12$ dots.

A third way to find the total number of dots is to skip count by 4s: 4, 8, 12. $3 \times 4 = 12$ is a multiplication number sentence for the problem. The × is a multiplication sign. You can read 3×4 as "3 groups of 4." You can also read it as "3 times 4" or as "3 multiplied by 4."

There are many strategies you can use to find the total amount in a set of equal groups. Some strategies are more **efficient,** or faster and easier, than others.

Equal groups of birds' eggs

Division: Equal Groups

You can use what you know about multiplication and division to solve number stories about equal groups.

The division sign (÷) can be used to show division in number models for equal-sharing and equal-grouping stories.

In an **equal-grouping** number story, you take a set of objects, make groups with the same number of objects in each group, and find the number of groups.

Example

Twenty children want to play ball. How many teams can be made with 5 children on each team?

Each team of 5 children is one group. You can use counters, tallies, or drawings to find how many teams of 5 can be made with 20 children.

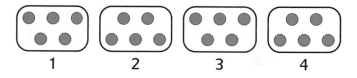

The counters show that 20 children can be divided into 4 equal groups. So 4 teams can be made.

A number model for this number story is 20 ÷ 5 = 4. Read this as "20 divided by 5 equals 4." It means that 20 children divided into teams of 5 make 4 teams.
Another number model for this story is ＿＿ × 5 = 20.
Think: What number times 5 equals 20? 4 × 5 = 20
This number model means 4 teams of children with 5 children on each team make 20 children in all.

Sometimes it is not possible to divide a set into equal groups without some left over. The number left over is called the **remainder.**

Example

23 children want to play ball. How many teams can be made with 5 children on each team?

The tallies to the right show that 23 children can be divided into 4 teams, with 3 children left over. The remainder is 3.

A number model for this number story is 23 ÷ 5 → 4 remainder 3, or 4 R3.

卌
卌
卌
卌
///

Division: Equal Shares

An **equal-sharing** number story involves dividing a set of things into equal parts and finding the number of things in each part. Equal parts are also called equal shares.

Example

Four boys share 24 marbles equally. How many marbles does each boy get?

You know the number of groups is 4. You need to find what share of the marbles is in each group. This is an equal-sharing story.

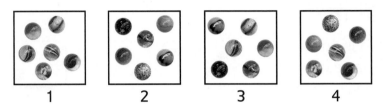

1   2   3   4

You can divide the 24 marbles into 4 equal groups.

A number model for the problem is $24 \div 4 = 6$.

Another number model for the problem is $4 \times \underline{\quad} = 24$. $4 \times 6 = 24$

Each boy's share is 6 marbles.

Example

Four boys share 26 marbles equally.

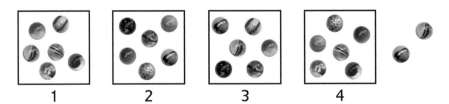

1   2   3   4

A number model for this problem is $26 \div 4 \rightarrow 6$ remainder 2, or 6 R2.

Each share has 6 marbles. The remainder is 2 since 2 marbles are left over.

University of Chicago

Arrays

An **array** is a group of objects arranged in equal **rows** and **columns.**

- Each row is filled and has the same number of objects.
- Each column is filled and has the same number of objects.
- If you draw an outline around the rows and columns, the outline will be a rectangle.

Example

The eggs in this carton form an array.

Here are ways to describe the array:

- It has 2 rows, with 6 eggs in each row.
- It is a 2-by-6 array.

Example

The keys on a phone keypad are arranged in an array.

The red outline around the keys is a rectangle.

There are 4 rows of 3 keys each.

This is called a 4-by-3 array.

Example

There are four different ways to make an array that has ten objects.

2-by-5 array   5-by-2 array   10-by-1 array

1-by-10 array

Arrays are useful for showing **equal groups** of objects. Each row in the array shows one of the equal groups.

Example

Louise buys 4 packages of juice boxes. Each package has 6 juice boxes. Show the 4 packages as an array.

There are 4 groups (4 packages of juice boxes).

Each group has the same number of objects (6 juice boxes).

Draw a 4-by-6 array to show the 4 packages.

Each row stands for 1 package of 6 juice boxes.

Drawing or building arrays can help you solve problems involving equal groups.

Example

There are 6 rows with 4 tomato plants in each row. How many tomato plants are there in all?

The 6 rows are the 6 groups. Each row has the same number of plants (4). So this is an equal-groups problem.

You can draw an array to show the 6 equal groups. Each row has 4 plants.

Multiply the number of rows by the number of objects in each row to find the total number of objects in the array.

Write 6×4 to show 6 groups of 4.

A number model for the problem is $6 \times 4 = 24$.

There are 24 plants in all.

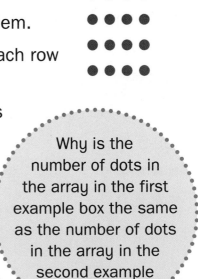

Why is the number of dots in the array in the first example box the same as the number of dots in the array in the second example box?

Arrays can also help you solve equal-sharing stories.

Example

There are 24 chairs. They are set up in 3 rows. How many chairs are in each row?

Arrange 24 counters in 3 rows.

There are 8 counters in each row.

Write 24 ÷ 3 = 8.

There are 8 chairs in each row.

You can also think: 3 times what number equals 24? Write 3 × 8 = 24.

Using either method, there are 8 chairs in each row.

Check Your Understanding

1. Write a number sentence for this array.

2. Write a number sentence for this array.

3. Elizabeth's garden has 32 plants. There are 4 rows of plants. How many plants are in each row? Draw an array and write a number model to match the story.

Check your answers in the Answer Key.

Multiplication Helper Facts

Helper facts are facts you know well that you can use to figure out facts that you don't know. The facts below are facts that you may already know well. They can help you solve other multiplication and division facts.

2s facts	Thinking of addition doubles can help you to solve 2s facts.	$2 \times 7 = ?$ *Think:* I have 2 groups of 7. I can write that as $7 + 7$. I know my doubles well, so I know $2 \times 7 = 14$.
10s facts	When you multiply by 10, you can think of having groups of 10. Then you can skip count by 10s easily.	$8 \times 10 = ?$ *Think:* I have 8 groups of 10. I can skip count by 10s until I have counted 8 times: 10, 20, 30, 40, 50, 60, 70, 80. So, $8 \times 10 = 80$.
5s facts	You can skip count by 5s to solve a 5s fact.	$5 \times 3 = ?$ *Think:* I can skip count by 5s until I have counted 3 times; 5, 10, 15. So, $5 \times 3 = 15$.
	You can also start with 10 groups of the number and take half of that to find 5 groups of the number.	$5 \times 8 = ?$ *Think:* I have 5 groups of 8. I know that 10 groups of 8 is 80. I can take half of that since I have half (5) of the number of groups (10). Half of 80 is 40, so $5 \times 8 = 40$.

Squares facts	When the factors are the same, you can imagine a square array since the number of rows and columns is the same.	$4 \times 4 = ?$
		 Think: I can draw a 4-by-4 array and count by 4s four times: 4, 8, 12, 16. So, $4 \times 4 = 16$.
		       × × × ×
		       × × × ×
		       × × × ×
		       × × × ×

Multiplication Facts Strategies

You can use facts strategies to become fluent with multiplication facts.

Turn-Around Rule for Multiplication

The turn-around rule for multiplication says you can multiply factors in any order. One way to show how the turn-around rule works is to turn an array. The number of rows becomes the number of columns, and the number of columns becomes the number of rows. Either way, the product is the same. Nothing is added to or taken away from the total amount.

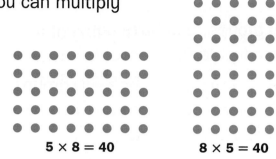

$5 \times 8 = 40$ $8 \times 5 = 40$

You can think about the turn-around rule in everyday equal-grouping situations.

Example

Lindsay has 8 boxes of 2 pens. How many pens does she have?

$8 \times 2 = 16$

She has 16 pens.

Tommy has 2 boxes of 8 pens. How many pens does he have?

$2 \times 8 = 16$

He has 16 pens.

Both Lindsay and Tommy have 16 pens.

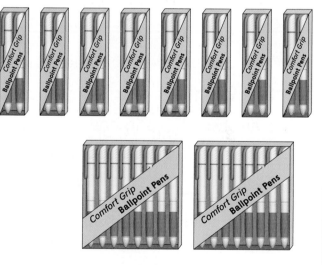

You can use the turn-around rule if changing the order of the factors makes a fact easier for you to solve. For example, since $5 \times 9 = 45$, you know that $9 \times 5 = 45$.

Multiplying by 0 and Multiplying by 1

When you multiply, you can think of making equal groups.

When you **multiply a number by 0,** you can think of having 0 groups of that number of objects. If you have 0 groups, you have nothing, so the product is 0.

0×5 means you have 0 groups of 5 objects, which means you have 0 objects in all.

When you **multiply a number by 1,** you can think of having 1 group of that number of objects. So the product will be equal to the number you multiplied by 1.

1×5 means you have 1 group of 5 objects, which means you have 5 objects in all.

Think about these number stories to make sense of multiplying by 0 and 1.

Example

Here is one pair of shoes. A pair is 1 group of 2 shoes.
How many shoes are in a pair?

$1 \times 2 = 2$

There are two shoes in one pair.

Example

Bagels are sold 12 in a bag. Sam purchases no bags of bagels.
How many bagels does he have?

$0 \times 12 = 0$

He has no bagels.

Multiplying More Than Two Factors

When you multiply more than two factors, it doesn't matter which two factors you multiply first. The product will be the same.

- $3 \times 5 \times 2 = ?$
- Start with $3 \times 5 = 15$.
- $15 \times 2 = 15 + 15 = 30$
- The product is 30.

- $3 \times 5 \times 2 = ?$
- Start with $5 \times 2 = 10$.
- $3 \times 10 = 30$
- The product is still 30.

Adding-a-Group Strategy

Adding a group to a helper fact can help you figure out other facts. You can show the **adding-a-group strategy** by adding a row to an array to find the new, larger product.

Example

$6 \times 4 = ?$

You know the helper fact $5 \times 4 = 20$.

Add another group of 4. $20 + 4 = 24$

So, $6 \times 4 = $ **24.**

```
× × × ×
× × × ×
× × × ×
× × × ×
× × × ×
× × × ×
```

To solve 6 × 4, you can start with a 5-by-4 array and add a row to make a 6-by-4 array.

The adding-a-group strategy works well for using mental math to figure out 3s facts and 6s facts because you can use 2s facts and 5s facts as helper facts.

Example

$3 \times 8 = ?$

Think: 3×8 is 3 groups of 8. That is close to 2 groups of 8, and I know $2 \times 8 = 16$.

I need to add 1 more group of 8.

Since $16 + 8 = 24$, I know $3 \times 8 = $ **24.**

```
× × × × × × × ×
× × × × × × × ×
× × × × × × × ×
```

To solve 3 × 8, you can imagine a 2-by-8 array and adding a row to make a 3-by-8 array.

Subtracting-a-Group Strategy

You can start from a helper fact and subtract a group to figure out a nearby fact. You can show the **subtracting-a-group strategy** by crossing out a row on an array to find the new, smaller product.

The subtracting-a-group strategy works well for using mental math to figure out 9s facts and 4s facts because you can use 10s facts and 5s facts as helper facts.

Example

$4 \times 6 = $ **?**

$5 \times 6 = 30$

$30 - 6 = 24$

$4 \times 6 = $ **24**

× × × × × ×
× × × × × ×
× × × × × ×
× × × × × ×
~~× × × × × ×~~

To solve 4 × 6, you can start with a 5-by-6 array and cross out a row.

Example

$9 \times 7 = $ **?**

Think: 9 × 7 is 9 groups of 7. That is close to 10 groups of 7, and I know 10 × 7 = 70.

I need to subtract the extra group of 7.

Since 70 − 7 = 63, I know 9 × 7 = **63.**

× × × × × × ×
× × × × × × ×
× × × × × × ×
× × × × × × ×
× × × × × × ×
× × × × × × ×
× × × × × × ×
× × × × × × ×
× × × × × × ×
~~× × × × × × ×~~

To solve 9 × 7, you can imagine a 10-by-7 array and crossing out a row to make a 9-by-7 array.

Check Your Understanding

Solve. Explain how to use the adding-a-group or subtracting-a-group strategy to solve Problems 1 and 3.

1. $3 \times 9 = $?   **2.** $6 \times 8 = $?   **3.** $9 \times 8 = $?   **4.** $4 \times 8 = $?

Check your answers in the Answer Key.

Doubling Strategy

Doubling can be helpful when the fact you are trying to solve is a double of a fact you know well. Area models can help you represent doubling.

Example

6 × 7 = **?**

Think of a rectangle with side lengths of 6 feet and 7 feet.

To find the area of the whole rectangle,

- break the rectangle into two halves that are each 3-by-7 feet,
- find the area of one of the rectangles,
- and double that area to solve 6 × 7.

21 + 21 = 42, so 6 × 7 = **42.**

7 ft

3 ft	3 × 7
3 ft	3 × 7

6 ft

3 × 7 = 21
3 × 7 = 21

21 + 21 = 42
so 6 × 7 = **42** sq ft

The area of the 6-by-7 rectangle is 42 square feet.
The product of 6 × 7 is 42.

Doubling works for facts with at least one even factor because you can start by halving that factor.

Example

4 × 7 = **?**

Think: I know half of 4 is 2, so I start with 2 × 7 = 14.

Then I can double 2 × 7 to find 4 × 7.

14 + 14 = 28,

so 4 × 7 = **28.**

7

2	2 × 7
2	2 × 7

4

2 × 7 = 14
2 × 7 = 14

14 + 14 = 28
so 4 × 7 = **28**

To solve 4 × 7, you can imagine a 2 × 7 rectangle and double the area.

Near-Squares Strategy

You can solve facts that are close to a multiplication square by using the **near-squares strategy.** You can add or subtract one or more groups from a nearby square to find the product.

```
× × × × × × ×
× × × × × × ×
× × × × × × ×
× × × × × × ×
× × × × × × ×
× × × × × × ×
× × × × × × ×
    7 × 7 = 49
```

```
× × × × × × ×
× × × × × × ×
× × × × × × ×
× × × × × × ×
× × × × × × ×
× × × × × × ×
× × × × × × ×
    6 × 7 = 42
```

```
× × × × × × ×
× × × × × × ×
× × × × × × ×
× × × × × × ×
× × × × × × ×
× × × × × × ×
× × × × × × ×
● ● ● ● ● ● ●
    8 × 7 = 56
```

You can start with 7 × 7 = 49 as your helper fact.

You can subtract a group from 7 × 7 to solve 6 × 7.

You can add a group to 7 × 7 to solve 8 × 7.

The near-squares strategy works well when the factors are close together.

Example

$9 \times 8 = ?$

Think: I know that 9×8 is close to 8×8 and that $8 \times 8 = 64$.

So I start with 8 groups of 8, but I need 9 groups of 8.

I need to add another group of 8.

Since $64 + 8 = 72$, I can find that $9 \times 8 = \textbf{72.}$

```
× × × × × × × ×
× × × × × × × ×
× × × × × × × ×
× × × × × × × ×
× × × × × × × ×
× × × × × × × ×
× × × × × × × ×
× × × × × × × ×
● ● ● ● ● ● ● ●
    9 × 8 = 72
```

To solve 9 × 8, you can imagine an 8-by-8 array and adding a row to make a 9-by-8 array.

Check Your Understanding

Solve. Explain how to use the doubling strategy or the near-squares strategy for Problems 2 and 3.

1. 4×9 **2.** 8×6 **3.** 5×6

Check your answers in the Answer Key.

Break-Apart Strategy

You can solve a fact by using the **break-apart strategy.** First break apart the fact to make two smaller helper facts. Then add the products of the smaller facts to find the product of the original fact. Area models can help you keep track of your steps.

Example

$7 \times 8 = ?$

Think of a rectangle with side lengths of 7 feet and 8 feet.

Since 5s and 2s facts can be helper facts, break apart 7 feet into 5 feet and 2 feet. The smaller rectangles have areas that are easier to find.

$7 \times 8 = 56$

8 ft

| | 5 ft | $5 \times 8 = 40$ |
7 ft

2 ft | $2 \times 8 = 16$

$40 + 16 = 56$

So $7 \times 8 = 5 \times 8 + 2 \times 8$
$= 40 + 16$
$= 56$ sq ft

The area of the 7-by-8 rectangle is 56 square feet. The product of 7×8 is 56.

When you use the break-apart strategy, start by choosing a factor to break apart.

Example

$9 \times 6 = ?$

Think: I'll break 9 into 5 and 4 since I know my 5s facts and some of my 4s facts.

9×6 breaks into $5 \times 6 + 4 \times 6$.

$5 \times 6 = 30$, and $4 \times 6 = 24$. That means $9 \times 6 = 30 + 24 = 54$.

So, $9 \times 6 = 54.$

6

5 | $5 \times 6 = 30$
9
4 | $4 \times 6 = 24$

$30 + 24 = 54$

To solve 9×6, you can imagine a 9-by-6 rectangle broken into smaller rectangles. Then you can use helper facts to find the smaller products.

Properties of Multiplication

Certain facts, or properties, are true for all numbers. For example, when you multiply 0 by any number, the product is always 0. You use properties of multiplication when you solve multiplication facts and apply multiplication strategies. The table below summarizes some important multiplication strategies and properties.

Multiplication Strategy	Examples
Multiplying by 1 When you multiply a number by 1, the product is that number.	$1 \times 3 = 3$ $48 \times 1 = 48$
Multiplying by 0 When you multiply a number by 0, the product is 0.	$0 \times 2 = 0$ $312 \times 0 = 0$
Turn-Around Rule When you multiply two numbers, the order doesn't matter.	$3 \times 4 = 12 \quad 4 \times 3 = 12$ $10 \times 4 = 40 \quad 4 \times 10 = 40$
Multiplying More Than Two Factors When you multiply three numbers, it doesn't matter which two numbers you multiply first.	$4 \times 2 \times 5 = ?$ or $4 \times 2 \times 5 = ?$ $8 \times 5 = \mathbf{40}$ $4 \times 10 = \mathbf{40}$
Break-Apart Strategy	9×8, breaking apart 9 into 5 and 4 $9 \times 8 = 5 \times 8 + 4 \times 8$ $9 \times 8 = 40 + 32$ $9 \times 8 = 72$

Check Your Understanding

1. $79 \times 1 = ?$ **2.** $0 \times 79 = ?$

3. Solve $7 \times 4 \times 5$ in two different ways. Which way was easier? Why?

4. Solve 6×8 using the break-apart strategy.

Check your answers in the Answer Key.

Fact Triangles and Fact Families

Fact Triangles are tools that can help you learn basic math facts. Below is a Fact Triangle card. The "×, ÷" printed on the card means that it is used to practice multiplication and division facts. The number in the • corner is the product of the other two numbers.

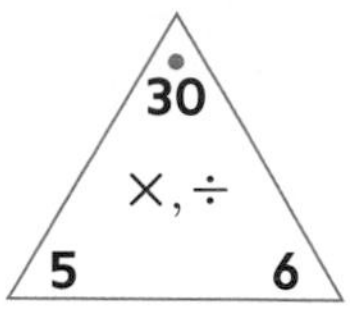

The **fact family** for this Fact Triangle is

$$5 \times 6 = 30 \qquad 30 \div 5 = 6$$
$$6 \times 5 = 30 \qquad 30 \div 6 = 5$$

You can work with a partner when you use Fact Triangles to practice basic facts. One partner covers one of the three corners with a finger. The other partner gives a complete multiplication or division fact.

Example

Here are ways to use the Fact Triangle shown above.

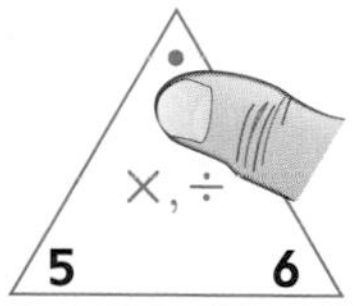

Lara covers 30. Denise says "5 × 6 equals 30" or "6 × 5 equals 30."

Lara covers 5. Denise says "30 ÷ 6 equals 5."

Lara covers 6. Denise says "30 ÷ 5 equals 6."

Check Your Understanding

1. Write the fact family for the Fact Triangle shown.

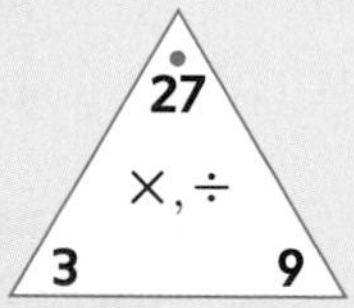

2. Draw a Fact Triangle for the fact family below.
   Write the three numbers of the fact family on the Fact Triangle.

   $$7 \times 8 = 56 \qquad 56 \div 7 = 8$$
   $$8 \times 7 = 56 \qquad 56 \div 8 = 7$$

3. Use your Fact Triangles to practice the multiplication facts. Put all the multiplication facts you know well in one pile. Think of strategies you can use to solve the rest of the facts and practice solving them.

Check your answers in the Answer Key.

Multiplication and Division Facts Table

A **facts table** is a chart with rows and columns. The factors 1 through 10 are listed along the outside edges in the left column and in the top row. The products of the multiplication facts 1 through 10 fill the rest of the table. You can find interesting patterns in the Multiplication/Division Facts Table.

×,÷	1	2	3	4	5	6	7	8	9	10
1	1	2	3	4	5	6	7	8	9	10
2	2	4	6	8	10	12	14	16	18	20
3	3	6	9	12	15	18	21	24	27	30
4	4	8	12	16	20	24	28	32	36	40
5	5	10	15	20	25	30	35	40	45	50
6	6	12	18	24	30	36	42	48	54	60
7	7	14	21	28	35	42	49	56	63	70
8	8	16	24	32	40	48	56	64	72	80
9	9	18	27	36	45	54	63	72	81	90
10	10	20	30	40	50	60	70	80	90	100

You can find fact families in the table.

Example

What fact family can you find where the 4-row and the 6-column meet?

Go across the 4-row to the 6-column. This row and column meet at a square that shows the number 24.

You can use the numbers 4, 6, and 24 to write the fact family made up of two multiplication facts and two division facts:

$4 \times 6 = 24$ $24 \div 4 = 6$

$6 \times 4 = 24$ $24 \div 6 = 4$

6-column

×,÷	1	2	3	4	5	6	7	8	9	10
1	1	2	3	4	5	6	7	8	9	10
2	2	4	6	8	10	12	14	16	18	20
3	3	6	9	12	15	18	21	24	27	30
4	4	8	12	16	20	24	28	32	36	40
5	5	10	15	20	25	30	35	40	45	50
6	6	12	18	24	30	36	42	48	54	60
7	7	14	21	28	35	42	49	56	63	70
8	8	16	24	32	40	48	56	64	72	80
9	9	18	27	36	45	54	63	72	81	90
10	10	20	30	40	50	60	70	80	90	100

4-row →

You can see the **turn-around rule** in the table.

Example

Find the product of 4 × 9.

- Move across the 4-row.
- Move down the 9-column.
- Stop where the row and column meet.

4 × 9 = 36

Find the product of 9 × 4.

- Move across the 9-row.
- Move down the 4-column.
- Stop where the row and column meet.

9 × 4 = 36

×,÷	1	2	3	4	5	6	7	8	9	10
1	1	2	3	4	5	6	7	8	9	10
2	2	4	6	8	10	12	14	16	18	20
3	3	6	9	12	15	18	21	24	27	30
4	4	8	12	16	20	24	28	32	36	40
5	5	10	15	20	25	30	35	40	45	50
6	6	12	18	24	30	36	42	48	54	60
7	7	14	21	28	35	42	49	56	63	70
8	8	16	24	32	40	48	56	64	72	80
9	9	18	27	36	45	54	63	72	81	90
10	10	20	30	40	50	60	70	80	90	100

You can use patterns in the table to help you find facts quickly.

Example

What pattern do square products make?

Locate the products of 1 × 1, 2 × 2, 3 × 3, and 4 × 4 on the table.

The products form a diagonal line on the table.

Using the diagonal pattern, you can predict the product of 5 × 5 without moving across the 5-row and down the 5-column. The product of 5 × 5 is 25.

×,÷	1	2	3	4	5	6	7	8	9	10
1	1	2	3	4	5	6	7	8	9	10
2	2	4	6	8	10	12	14	16	18	20
3	3	6	9	12	15	18	21	24	27	30
4	4	8	12	16	20	24	28	32	36	40
5	5	10	15	20	25	30	35	40	45	50
6	6	12	18	24	30	36	42	48	54	60
7	7	14	21	28	35	42	49	56	63	70
8	8	16	24	32	40	48	56	64	72	80
9	9	18	27	36	45	54	63	72	81	90
10	10	20	30	40	50	60	70	80	90	100

Example

What patterns do the products in the 9-column make?

- The tens digit counts up by 1 going down the 9-column.
- The ones digit counts down by 1 going down the 9-column.
- If you add the digits together, the sum is always 9. For example, $1 + 8 = 9$ and $2 + 7 = 9$.

×,÷	1	2	3	4	5	6	7	8	9	10
1	1	2	3	4	5	6	7	8	9	10
2	2	4	6	8	10	12	14	16	18	20
3	3	6	9	12	15	18	21	24	27	30
4	4	8	12	16	20	24	28	32	36	40
5	5	10	15	20	25	30	35	40	45	50
6	6	12	18	24	30	36	42	48	54	60
7	7	14	21	28	35	42	49	56	63	70
8	8	16	24	32	40	48	56	64	72	80
9	9	18	27	36	45	54	63	72	81	90
10	10	20	30	40	50	60	70	80	90	100

Check Your Understanding

What patterns do you notice when you compare the products of 5s facts and 10s facts?

×,÷	1	2	3	4	5	6	7	8	9	10
1	1	2	3	4	5	6	7	8	9	10
2	2	4	6	8	10	12	14	16	18	20
3	3	6	9	12	15	18	21	24	27	30
4	4	8	12	16	20	24	28	32	36	40
5	5	10	15	20	25	30	35	40	45	50
6	6	12	18	24	30	36	42	48	54	60
7	7	14	21	28	35	42	49	56	63	70
8	8	16	24	32	40	48	56	64	72	80
9	9	18	27	36	45	54	63	72	81	90
10	10	20	30	40	50	60	70	80	90	100

Check your answers in the Answer Key.

Extended Facts

Basic facts can help you solve extended facts. **Extended facts** are problems that include a multiple of 10. Related basic multiplication facts can help you find the number of tens (or hundreds or thousands) in a solution.

Example

$5 \times 30 = $ **?**

Think: I know the basic fact $5 \times 3 = 15$.
Since 30, not 3, is in the problem, I can think about 30 as 3 groups of 10, or 3 tens.

5×3 tens $= 15$ tens

So, $5 \times 30 = 150$.

$5 \times 3 = 15$ 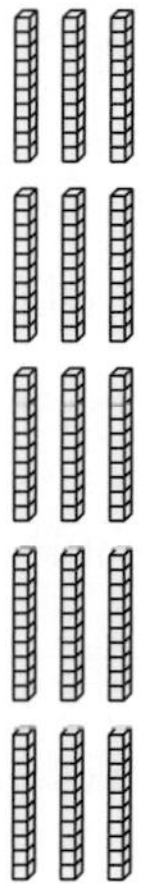 $5 \times 30 = 150$

Knowing the basic fact $5 \times 3 = 15$ can help you solve the related extended fact $5 \times 30 = $ **150.**

Example

$2 \times 600 = $ **?**

Think: 600 is 6 groups of one hundred, or 6 [100s].

2×6 [100s] $= $?

The basic fact $2 \times 6 = 12$ is related to the problem.

2×6 [100s] $= 12$ [100s]

So, $2 \times 600 = $ **1,200.**

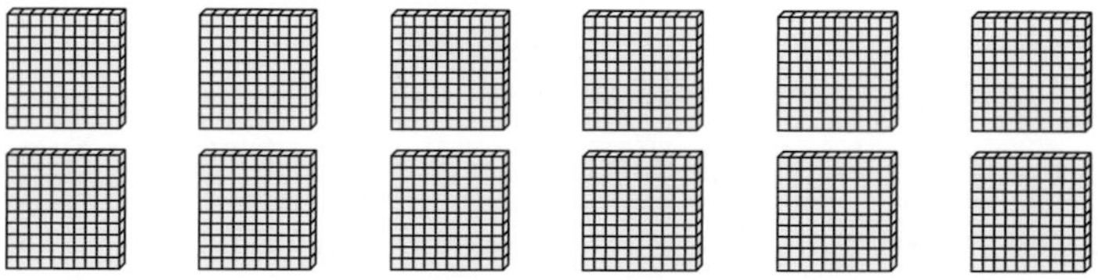

6 groups of 2 hundreds make 12 hundreds in all.

Your knowledge of basic facts can help you solve division problems with multiples of 10.

Example

$160 \div 8 = \mathbf{?}$

Think: 160 is 16 groups of ten, or 16 [10s].

$16 \text{ [10s]} \div 8 = ?$

The basic fact $16 \div 8 = 2$ is related to the problem.

$16 \text{ [10s]} \div 8 = 2 \text{ [10s]}$

So, $160 \div 8 = \mathbf{20.}$

16 tens divided into 8 groups means 2 tens are in each group.

Example

$200 \div 5 = \mathbf{?}$

Think: 200 is 20 groups of ten, or 20 [10s].

$20 \text{ [10s]} \div 5 = ?$

The basic fact $20 \div 5 = 4$ is related to the problem.

$20 \text{ [10s]} \div 5 = 4 \text{ [10s]}$

So, $200 \div 5 = \mathbf{40.}$

You can imagine 20 tens divided into 5 groups so that 4 tens are in each group.

Check Your Understanding

Record a basic multiplication fact that could help you solve the problem. Then solve.

1. $7 \times 40 = ?$ **2.** $? = 800 \times 3$ **3.** $360 \div 4 = ?$ **4.** $? = 300 \div 6$

Check your answers in the Answer Key.

Multiplication Strategies for Larger Factors

You can multiply larger numbers by using strategies you already know for finding products of basic multiplication facts.

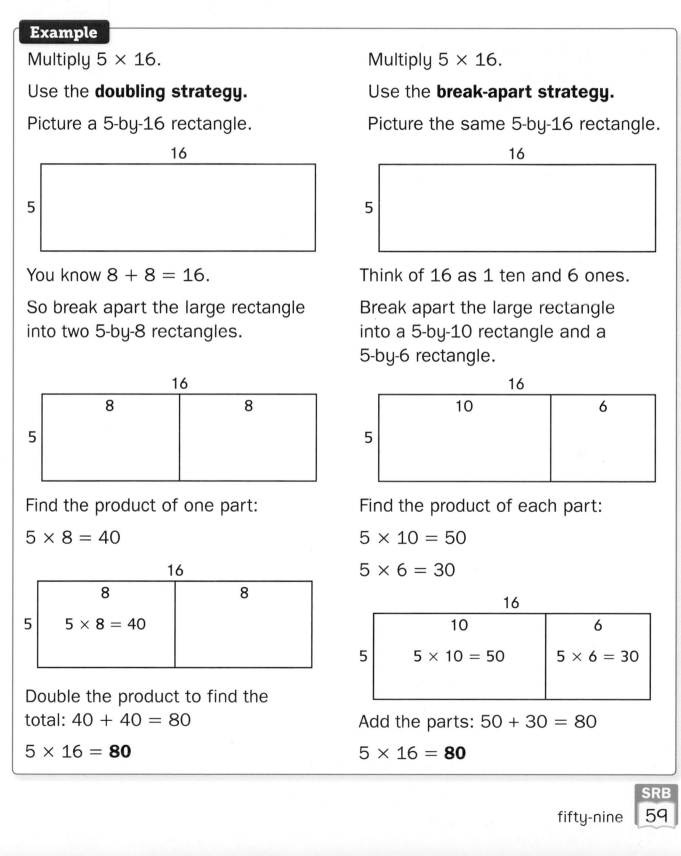

Example

Multiply 5 × 16.

Use the **doubling strategy.**

Picture a 5-by-16 rectangle.

16
5

You know 8 + 8 = 16.

So break apart the large rectangle into two 5-by-8 rectangles.

16
8   8
5

Find the product of one part:

5 × 8 = 40

16
8   8
5   5 × 8 = 40

Double the product to find the total: 40 + 40 = 80

5 × 16 = **80**

Multiply 5 × 16.

Use the **break-apart strategy.**

Picture the same 5-by-16 rectangle.

16
5

Think of 16 as 1 ten and 6 ones.

Break apart the large rectangle into a 5-by-10 rectangle and a 5-by-6 rectangle.

16
10   6
5

Find the product of each part:

5 × 10 = 50

5 × 6 = 30

16
10   6
5   5 × 10 = 50   5 × 6 = 30

Add the parts: 50 + 30 = 80

5 × 16 = **80**

Example

Multiply 42 × 7. Use the break-apart strategy.

Picture a 42-by-7 rectangle.

Think of 42 as 20 + 20 + 2.

Break apart the large rectangle into two
20-by-7 rectangles and a 2-by-7 rectangle.

Write a number sentence for each part of the whole rectangle.

20 × 7 = 140

20 × 7 = 140

2 × 7 = 14

Then add up all the parts to find the total.

140 + 140 + 14 = 294

42 × 7 = **294**

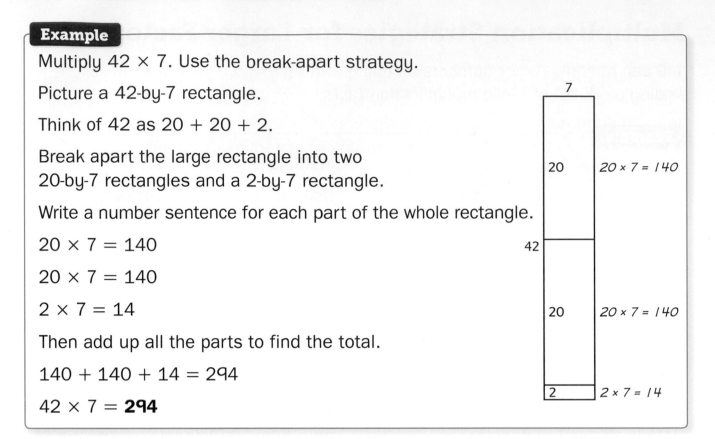

Example

Multiply 42 × 7. Use the break-apart strategy another way.

Picture a 42-by-7 rectangle.

Think of 42 as 4 tens (40) and 2 ones (2).

Break apart the large rectangle into a
40-by-7 rectangle and a 2-by-7 rectangle.

Write a number sentence representing each part of the
whole rectangle. Then add the parts.

42 × 7 = (40 × 7) + (2 × 7)

     =   280   +   14

     =   294

42 × 7 = **294**

The extended fact 40 × 7 is added to the basic fact
2 × 7 to find the total product.

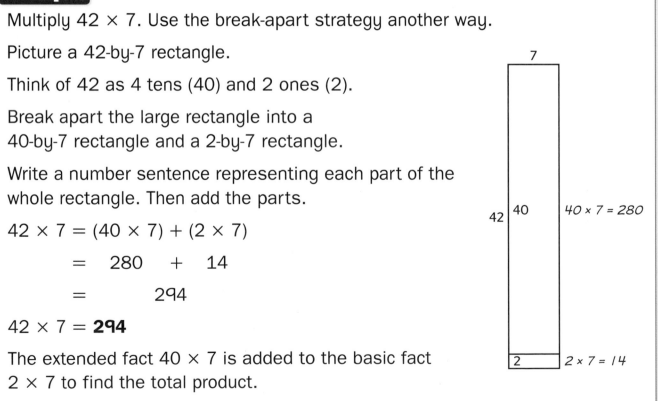

Representations for Unknown Values

Letters, blanks, and question marks can represent unknown values in number sentences and number models.

In the number sentence $5 + n = 8$, the unknown value is 3. You can replace the n with 3 to make the number sentence true. The summary number sentence is $5 + 3 = 8$.

Note A **number sentence** is a sentence that uses numbers and symbols. A **number model** is a number sentence that represents a situation in the real world.

Example

Here are number sentences with different symbols for unknown values.

Number Sentence	$4 + \_\_ = 15$	$8 \div ? = 2$	$25 = 10 + x$	$y = 6 \times 9$
Summary Number Sentence	$4 + 11 = 15$	$8 \div 4 = 2$	$25 = 10 + 15$	$54 = 6 \times 9$

You can use a letter to stand for the unknown amount in a number model for a number story. Choosing the first letter of the unit of the unknown amount can help you remember the unit you are trying to find.

Example

The temperature was 80 degrees at 6:00 A.M. At noon it was 100 degrees. What was the temperature change?

Use the letter t to stand for the temperature change.

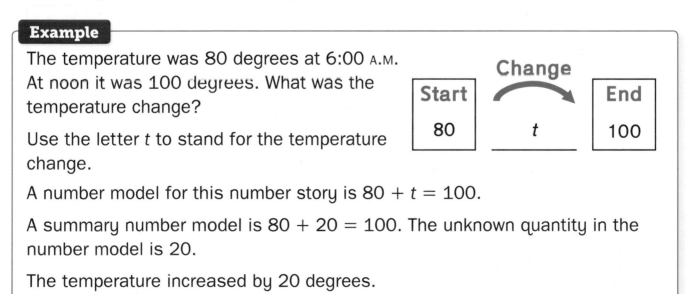

A number model for this number story is $80 + t = 100$.

A summary number model is $80 + 20 = 100$. The unknown quantity in the number model is 20.

The temperature increased by 20 degrees.

There is often more than one way to write a number model for a problem.

Example

There are 25 children in Mr. Lopez's class. Twelve children are girls. The rest are boys. How many are boys?

Use the letter *b* to stand for the number of boys.

One way:

A number model for this number story is $25 = 12 + b$.

A summary number model is $25 = 12 + 13$.
The unknown amount in the number model is 13.

Total children 25	
Part girls 12	**Part** boys *b*

Another way:

A number model for this number story is $25 - 12 = b$.

A summary number model is $25 - 12 = 13$. The unknown amount in the number model is 13.

There are 13 boys in the class.

Example

Eleanor's high score on a video game is 165 points more than Annie's high score. If Annie's high score is 574, what is Eleanor's high score?

A number model for this story is
$p = 574 + 165$.

A summary number model is
$739 = 574 + 165$.

The unknown quantity in the number model is 739.

Eleanor's high score is 739 points.

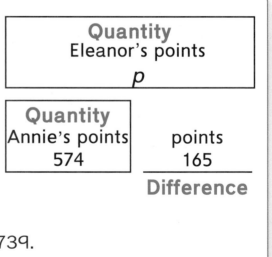

Quantity Eleanor's points
p

Quantity Annie's points	
574	points 165
	Difference

Multiplication/Division Diagram

You can use a diagram to help organize information from a number story. A **multiplication/division diagram** is useful for multiplication, equal-groups, and equal-shares problems. The diagram has spaces to keep track of three things:

- the number of groups
- the number of objects per group (**per** means "in each")
- the total number of objects

groups	objects per group	objects in all

First make sense of the number story. What do you know? What do you want to find out?

Fill in the diagram with the amounts you know. Then write a question mark (?) or a letter for the amount you want to find.

Example

9 stickers are on one strip of stickers. Tiffany has 8 strips of stickers. How many stickers does she have in all?

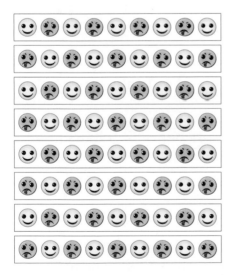

strips	stickers per strip	stickers in all
8	9	s

A number model for this problem is 8 × 9 = s.

A summary number model is 8 × 9 = 72.
The missing quantity in the number model is 72.

Tiffany has 72 stickers.

Sometimes you know the total number of objects but must find the number of groups or the number of objects in each group. You can multiply or divide.

Example

54 cards are placed in 6 equal piles. How many cards are in each pile?

You know the total number of objects and the number of piles (groups). You need to find the number of objects, or cards, in each pile.

piles	cards per pile	cards in all
6	c	54

You can divide 54 by 6. A number model for this story is $54 \div 6 = c$.

Or you can think: Six times what number equals 54? Another number model is $6 \times c = 54$.

Summary number models for this problem are $54 \div 6 = 9$ and $6 \times 9 = 54$. The missing quantity in each number model is 9.

There are 9 cards in each pile.

Example

Each table in a lunchroom must have exactly 6 chairs. There are 33 chairs. How many tables can have exactly 6 chairs?

You know the total number of objects and the number of objects per group. You need to find the number of groups.

Key: ▭ = 1 table • = 1 chair

You can divide the total number of chairs by the number of chairs at one table. Divide 33 by 6.

Or you can think: What number times 6 equals 33?

tables	chairs per table	chairs in all
t	6	33

$33 \div 6 \rightarrow 5$ R3 is a number model for this problem.

You can also write $5 \times 6 = 30$, $30 + 3 = 33$ to show how to solve this problem.

Five tables can have 6 chairs. There are 3 chairs left over.

Factors of a Number

When two numbers are multiplied, the answer is called the **product.** The two numbers that are multiplied are called **factors** of the product. These two factors together make a **factor pair.**

Note When you find the factors of a counting number, those factors must also be counting numbers. **Counting numbers** are the numbers 1, 2, 3, and so on.

Example

$3 \times 5 = 15$

15 is the product of 3 and 5. 3 and 5 are factors of 15. 3 and 5 make a factor pair of 15.

You can find factors of a number with arrays. If you can make an array with no leftovers, the number of rows and the number of columns make a factor pair.

Example

Find factor pairs of 6. Use counters to make arrays.

The counters can be arranged in 1 row with 6 counters in the row. 1 and 6 make a factor pair of 6.

The counters can be arranged in 2 equal rows with 3 counters in each row. They can also be arranged in 3 equal rows with 2 counters in each row. 2 and 3 make a factor pair of 6.

The counters cannot be arranged in 4 equal rows. 4 is *not* a factor of 6. 4 cannot be paired with another counting number to make a product of 6.

The counters cannot be arranged in 5 equal rows. 5 is *not* a factor of 6. 5 cannot be paired with another counting number to make a product of 6.

The factor pairs of 6:   1 and 6   2 and 3

The factors of 6 are 1, 2, 3, and 6.

Example

Find factor pairs of 9. Use counters to make arrays.

• • • • • • • • •

9 counters can be arranged in 1 row with 9 counters in the row. 1 and 9 make a factor pair of 9.

• • •
• • •
• • •

9 counters can be arranged in 3 equal rows with 3 counters in each row. 3 and 3 make a factor pair of 9.

Since you found that 1 and 9 make a factor pair, you do not need to show a 9-by-1 array. You know from the turn-around rule that both 1 × 9 and 9 × 1 work. No other arrangements of 9 counters make arrays with equal rows and equal columns.

The factor pairs of 9: 1 and 9   3 and 3
The factors of 9 are 1, 3, and 9.

You can also find the factors of a number by thinking about the multiplication facts for that product.

Example

Find the factor pairs of 24.

Multiply two counting numbers to get a product of 24. The factor pairs are shown in the table.

Factor Pairs of 24	
1 × 24	1 and 24
2 × 12	2 and 12
3 × 8	3 and 8
4 × 6	4 and 6

1, 2, 3, 4, 6, 8, 12, and 24 are the factors of 24.

Example

Find all the factor pairs of 7.

Multiply two counting numbers to get a product of 7: 1 × 7 = 7.
The only two counting numbers that you can multiply to get 7 are 1 and 7.

The only factor pair of 7 is 1 and 7.

The numbers 1 and 7 are all the factors of 7.

Factors of Prime and Composite Numbers

A counting number that has exactly two different factors is called a **prime number.**

A counting number that has three or more different factors is called a **composite number.**

The only way to multiply two counting numbers to get 1 is $1 \times 1 = 1$. So the number 1 has only one factor: 1. Since a prime number has exactly two different factors, 1 is not a prime number. Since a composite number has at least three different factors, 1 is not a composite number.

The table below shows different ways to list factors of a number and tells whether a number is prime, composite, or neither. Notice that the two factors of a prime number are always 1 and the number itself.

Number	Factor Pairs		Factors	Prime or Composite
1	1×1	1 and 1	1	neither
2	1×2	1 and 2	1, 2	prime
3	1×3	1 and 3	1, 3	prime
4	$1 \times 4, 2 \times 2$	1 and 4, 2 and 2	1, 2, 4	composite
5	1×5	1 and 5	1, 5	prime
6	$1 \times 6, 2 \times 3$	1 and 6, 2 and 3	1, 2, 3, 6	composite
7	1×7	1 and 7	1, 7	prime
8	$1 \times 8, 2 \times 4$	1 and 8, 2 and 4	1, 2, 4, 8	composite
9	$1 \times 9, 3 \times 3$	1 and 9, 3 and 3	1, 3, 9	composite
10	$1 \times 10, 2 \times 5$	1 and 10, 2 and 5	1, 2, 5, 10	composite
11	1×11	1 and 11	1, 11	prime
12	$1 \times 12, 2 \times 6, 3 \times 4$	1 and 12, 2 and 6, 3 and 4	1, 2, 3, 4, 6, 12	composite
13	1×13	1 and 13	1, 13	prime
14	$1 \times 14, 2 \times 7$	1 and 14, 2 and 7	1, 2, 7, 14	composite
15	$1 \times 15, 3 \times 5$	1 and 15, 3 and 5	1, 3, 5, 15	composite
16	$1 \times 16, 2 \times 8, 4 \times 4$	1 and 16, 2 and 8, 4 and 4	1, 2, 4, 8, 16	composite
17	1×17	1 and 17	1,17	prime
18	$1 \times 18, 2 \times 9, 3 \times 6$	1 and 18, 2 and 9, 3 and 6	1, 2, 3, 6, 9, 18	composite
19	1×19	1 and 19	1, 19	prime
20	$1 \times 20, 2 \times 10, 4 \times 5$	1 and 20, 2 and 10, 4 and 5	1, 2, 4, 5, 10, 20	composite

Parentheses

Parentheses ( ) are grouping symbols that help you understand number sentences with more than one operation. Parentheses make the meaning of a number sentence clear by showing which operation to do first.

Example

Solve. $9 - (3 + 2) = n$

The parentheses show that $3 + 2$ should be solved first.

Then subtract 5 from 9.

The answer is 4.

$9 - (3 + 2) = 4$

$$9 - (3 + 2) = n$$
$$9 - 5 \quad = n$$
$$4 \quad = n$$

Examples

Solve and compare.

$(15 - 3) + 2 = ?$	$15 - (3 + 2) = ?$
Subtract first. $15 - 3 = 12$	Add first. $3 + 2 = 5$
Then add. $12 + 2 = 14$	Then subtract. $15 - 5 = 10$
$(15 - 3) + 2 = \mathbf{14}$	$15 - (3 + 2) = \mathbf{10}$

Grouping the same numbers differently may give different answers.

You may need to include parentheses to make the number sentence true.

Example

Include parentheses to make this number sentence true: $14 - 6 \div 2 = 4$.

There are two possible places for the parentheses:

$(14 - 6) \div 2 = ?$	or	$14 - (6 \div 2) = ?$
Subtract first. $14 - 6 = 8$		Divide first. $6 \div 2 = 3$
Then divide. $8 \div 2 = 4$		Then subtract. $14 - 3 = 11$
$(14 - 6) \div 2 = 4$		$14 - (6 \div 2) = 11$
4 is equal to 4.		11 is not equal to 4.

The parentheses around $14 - 6$ make the number sentence true.

The correct place for the parentheses is $(14 - 6) \div 2 = 4$.

Order of Operations

To avoid confusion when solving number sentences, mathematicians have agreed to a set of rules called the **order of operations.** These rules tell you what to do first and what to do next.

> What is the answer when you add first in the number sentence $8 + 4 \times 3 =$ ____? What is the answer when you multiply first?

The rules are important to follow so that everyone gets the same answer to a problem. For example, think about $8 + 4 \times 3 =$ _____. If someone added first and someone else multiplied first, they would get different answers.

Rules for the Order of Operations
1. Do operations inside **parentheses** first.   Follow rules 2 and 3 when computing inside parentheses. 2. **Then multiply or divide,** in order, from left to right. 3. **Finally add or subtract,** in order, from left to right.

Example

Solve. $8 + 4 \times 3 = D$

Multiply first. $8 + 4 \times 3$

Then add. $= 8 + 12$

The answer is 20. $= 20$

$8 + 4 \times 3 = \mathbf{20}$

Example

Solve. _____ $= 10 - 6 + 2$

There are no parentheses and no multiplication or division signs, so follow Rule 3.

Rule 3 says to add and subtract in order from left to right.

Subtract first. $10 - 6 + 2$

Then add. $= 4 + 2$

The answer is 6. $= 6$

$\mathbf{6} = 10 - 6 + 2$

Example

Solve. $10 - (9 - 6 \div 2) = f$

Start inside parentheses.

Divide first. $10 - (9 - 6 \div 2)$

$= 10 - (9 - 3)$

Then subtract $9 - 3$. $10 - (9 - 3)$

$= 10 - 6$

Subtract $10 - 6$. $10 - 6$

The answer is 4. $= 4$

$10 - (9 - 6 \div 2) = \mathbf{4}$

Check Your Understanding

Solve. Be sure to follow the rules for the order of operations.

1. $8 \times (7 - 6) = L$

2. $j = 7 \times 3 - 21$

3. $(10 - 2 \times 4) \div 2 = B$

Check your answers in the Answer Key.

Note See page 296 in the Appendix to learn about how to use calculators to follow the order of operations.

Number Patterns

Dot patterns can be used to represent numbers. The dot pictures can help you find number patterns.

Note All of the dot pictures shown here are for counting numbers. The **counting numbers** are 1, 2, 3, and so on.

Even numbers

2	4	6	8	10	12

Even numbers are counting numbers that can be divided by 2 with no remainders. You can make an array with 2 equal rows for any even number.

Odd Numbers

1	3	5	7	9	11	13

Odd numbers are counting numbers that have a remainder of 1 when they are divided by 2. An odd number has a dot picture with 2 equal rows, plus 1 extra dot. You *cannot* make an array with 2 equal rows for an odd number.

Square Numbers

1	4	9	16	25

A **square number** is the product of a counting number multiplied by itself. For example, 16 is a square number because 16 equals 4 × 4. A square number has a dot picture that is an array with a square shape, with the same number of dots in each row and column.

Prime Numbers

2	3	5	7	11	13	17	19

A **prime number** is a counting number greater than 1 that has exactly 2 different factors, both of which are counting numbers. The dot picture for a prime number cannot fit into a rectangular shape (with at least 2 rows and at least 2 columns).

Frames and Arrows

A Frames-and-Arrows diagram is one way to show a number pattern. This type of diagram has three parts:

- a set of **frames** that contain numbers;
- **arrows** that show the path from one frame to the next frame;
- a box with an arrow and a **rule.** The rule tells how to change the number in one frame to get the number in the next frame.

Here is a Frames-and-Arrows diagram.

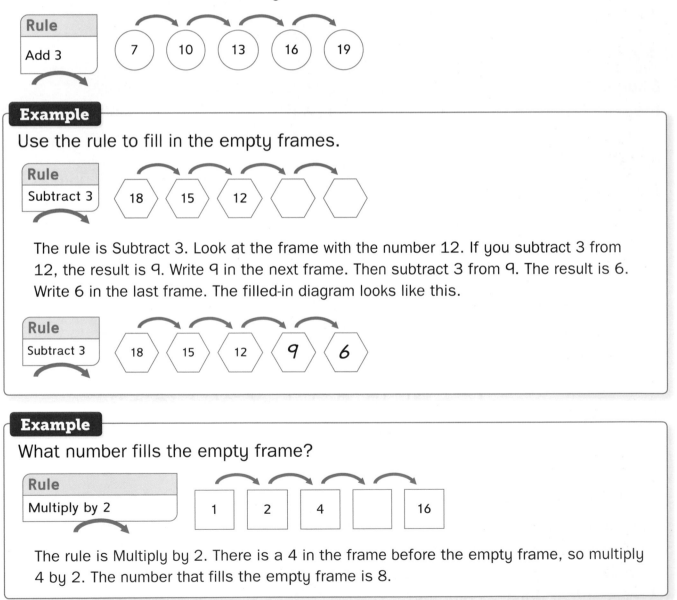

Rule

Add 3

7 10 13 16 19

Example

Use the rule to fill in the empty frames.

Rule

Subtract 3

18 15 12

The rule is Subtract 3. Look at the frame with the number 12. If you subtract 3 from 12, the result is 9. Write 9 in the next frame. Then subtract 3 from 9. The result is 6. Write 6 in the last frame. The filled-in diagram looks like this.

Rule

Subtract 3

18 15 12 9 6

Example

What number fills the empty frame?

Rule

Multiply by 2

1 2 4 □ 16

The rule is Multiply by 2. There is a 4 in the frame before the empty frame, so multiply 4 by 2. The number that fills the empty frame is 8.

Sometimes the rule is not given. You can use the frames to find the rule.

Example

Find a rule for this diagram.

Each number can be found by multiplying the number in the frame before it by 3.

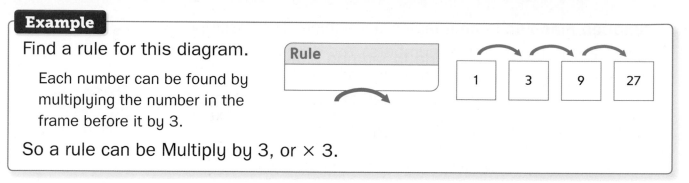

So a rule can be Multiply by 3, or × 3.

Sometimes the rule is not given and the frames are not all filled in. Find a rule first. Then use the rule to fill in the empty frames.

Example

Find a rule and fill in the empty frames.

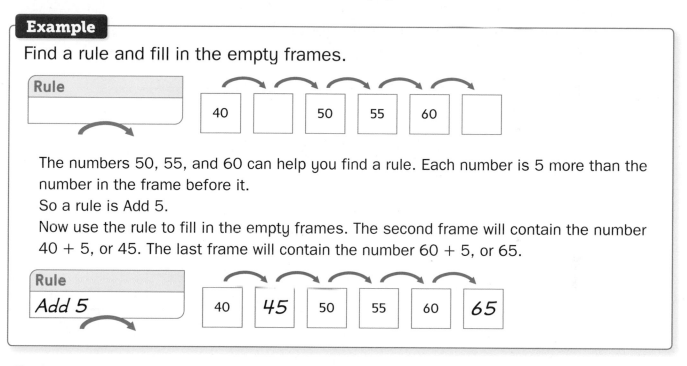

The numbers 50, 55, and 60 can help you find a rule. Each number is 5 more than the number in the frame before it.

So a rule is Add 5.

Now use the rule to fill in the empty frames. The second frame will contain the number 40 + 5, or 45. The last frame will contain the number 60 + 5, or 65.

Frames-and-Arrows diagrams can have more than one rule.

Example

Fill in the empty frames.
Follow the rules.

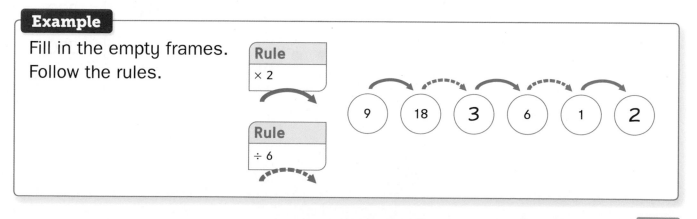

Function Machines and "What's My Rule?"

A **function machine** is an imaginary machine. The machine is given a rule for changing numbers. You drop a number into the machine. The machine uses the rule to change the number. The changed number comes out of the machine.

Here is a picture of a function machine.

The machine has been given the rule × 4.

The machine will multiply any number that is put into it by 4.

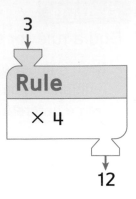

Example

If you drop 3 into the function machine above, it will multiply 3 × 4. The number 12 will come out.

If you drop 1 into the machine, it will multiply 1 × 4. The number 4 will come out.

If you drop 0 into the machine, it will multiply 0 × 4. The number 0 will come out.

You can use a table of *in* and *out* numbers to keep track of the way a function machine changes numbers.

Write the numbers that are put into the machine in the *in* column.

Write the numbers that come out of the machine in the *out* column.

in
↓
Rule
+ 150
↓
out

in	out
50	200
100	250
130	280
170	320

Example

The rule is × 10. You know the numbers that are put into the machine. Find the numbers that come out of the machine.

If 1 is put in, then 10 comes out. 1 × 10 = 10
If 3 is put in, then 30 comes out. 3 × 10 = 30
If 7 is put in, then 70 comes out. 7 × 10 = 70

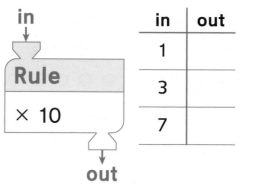

in	out
1	
3	
7	

Example

The rule is ÷ 2. You know the numbers that come out of the machine. Find the numbers that were put into the machine.

The machine divides any number put into it by 2. The number that comes out is always half the number that was put in.
If 9 comes out, then 18 was put in.
If 6 comes out, then 12 was put in.
If 4 comes out, then 8 was put in.

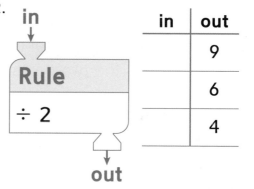

in	out
	9
	6
	4

Example

The rule is not known. Use the table to find a rule.

Each number in the *out* column is 9 times the number in the *in* column.

A rule can be Multiply by 9, or × 9.

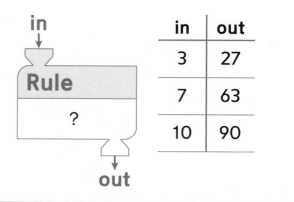

in	out
3	27
7	63
10	90

Diagrams for 1- and 2-Step Number Stories

You can solve number stories in many different ways. Drawings, arrays, and number lines are tools you can use to help you make sense of and solve a problem. Situation diagrams can also help you organize your thinking.

You can use diagrams for 1-step number stories.

Example

Rhodes Elementary School students donated 893 books in two weeks to a local community center. In the first week they donated 432 books. How many books did they donate in the second week?

Fill in a situation diagram. The letter B shows the unknown number of books donated in the second week.

Option 1: Think of this as a **parts-and-total** story where two or more parts are combined to form a total. A **parts-and-total diagram** shows that the total is the same size as the two parts combined.

Total	
893	
Part	**Part**
432	B

Option 2: Think of this as a **change-to-more** story. A **change diagram** has spaces to show the **Start, Change,** and **End** amounts in a number story.

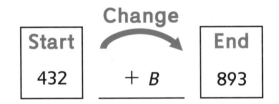

A number model for this story is $432 + B = 893$.
Another number model is $893 - 432 = B$.

Summary number models for this story are $432 + \textbf{461} = 893$ and $893 - 432 = \textbf{461}$.

461 books were donated in the second week.

You can use situation diagrams to organize your thinking for 2-step number stories.

Example

There are 3 bags of apples with 5 apples in each bag. There are also 10 oranges. How many pieces of fruit are there in all?

Anna thinks of the problem as a parts-and-total story. She writes *F*, for fruit, in the Total space.

Since one part of the total number of pieces of fruit is 3 bags of 5 apples, she writes (3 × 5) in one of the Part spaces.

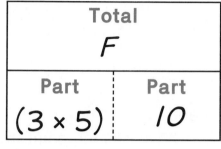

Another part of the total number of pieces of fruit is 10 oranges.
She writes 10 in the other Part space.

She writes a matching number model for this number story. $F = (3 × 5) + 10$

$F = (3 × 5) + 10$

$F = 15 + 10$

$F = 25$

A summary number model for the story is $25 = (3 × 5) + 10$.

There are 25 pieces of fruit in all.

Note Parentheses are symbols to show how to group numbers and operations in a number sentence. For example, in the problem above, the product in parentheses (3 × 5) describes the total number of apples in the problem.

You can also use a multiplication/division diagram to organize your thinking about 2-step number stories.

Miss Fry buys 6 packages of paints for her art classes. Each package has 4 paint trays. If she needs 8 trays for each class, how many classes will have enough paint?

She has 6 packages of 4 paint trays, or (6 × 4) paint trays.

She needs 8 paint trays per class.

She needs to find how many classes have enough paint trays.

classes	paint trays per class	paint trays in all
c	8	(6 × 4)

A number model for the story is $c \times 8 = (6 \times 4)$.

Think: 6 × 4 = 24

So, $c \times 8 = 24$.

Think: What number times 8 equals 24?

3 × 8 = 24

A summary number model for this number story is **3 × 8 = (6 × 4)**.

Miss Fry has enough paint trays for 3 classes.

Solve the problem. Organize the information in a situation diagram to help you write a number model for the story. Include a summary number model.

1. Miss Fry has 9 packages of 2 canvases. She needs 24 canvases for the children in her art class. How many more canvases does she need?

Check your answers in the Answer Key.

Mathematics in the Wild

Scientists study the world around us, including patterns in the lives of wild animals.

Migration Routes

Migration patterns are cycles of movement that many animals in the wild repeat each year.

Gray whales complete one of the longest migration routes of any mammal. In one year, a whale can travel about 10,000–14,000 miles.

Gray whales spend the summer months feeding in the cold waters of the Northern Pacific Ocean. Each fall, most of them swim south, traveling about 75 miles each day for 2–3 months. They breed and give birth in the warmer waters near Mexico, and then return north each spring.

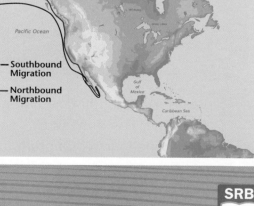

What other animals migrate?

Animal Growth

From birth to adulthood, the size of an animal increases in fairly predictable ways. Some animals, including many fish and reptiles, grow very quickly. They start out tiny and mature to be thousands of times their birth size.

A crocodile hatchling has a mass of about 60 grams. It weighs about 2 ounces.

A mature crocodile can have a mass of about 480 kilograms. It is about 8,000 times heavier than the hatchling.

A full-grown salmon can have a mass of more than 25 kilograms, making it about 100,000 times heavier than it was when it hatched.

When a salmon hatches from its egg, it has a mass of about one-fourth of a gram.

Other animals, including many mammals, start out larger and grow more slowly over time.

A newborn human infant has a mass of about 3 or 4 kilograms. The baby's mother may have a mass of about 64 kilograms. She is only about 18 times heavier than her baby.

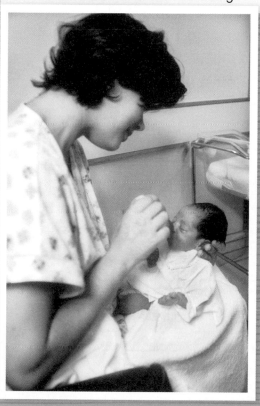

A newborn horse, or foal, has a mass of about 23 kilograms.

The foal's mother has a mass of about 700 kilograms. She is about 30 times heavier than her foal.

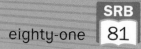

Animal Survival

All animals strive to reach adulthood, or maturity. The number of babies an animal has at a time is related to the number that is likely to survive. Many mammals and birds have 1 to 10 babies at a time, and their babies have a pretty good chance of surviving.

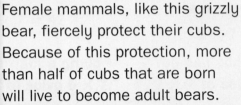

Female mammals, like this grizzly bear, fiercely protect their cubs. Because of this protection, more than half of cubs that are born will live to become adult bears.

Ducks lay about 5 to 15 eggs and keep them warm until they hatch.

This duck does her best to protect her ducklings. If all goes well, about 1 out of 4 of them will live to reach maturity.

Other animals, like reptiles, amphibians, and fish, lay many more eggs at a time, but their babies have very little chance of surviving to reach maturity.

Some frogs can lay 1,000 or more eggs. Fewer than 1 out of 100 of the tadpoles that hatch from the eggs will live to become adult frogs.

A crocodile lays up to about 90 eggs. Most of them never hatch because predators eat them. Fewer than 1 out of 10 of the hatched baby crocodiles will live to become adult crocodiles.

Many fish have even less chance of survival. The walleye lays more than 200,000 eggs. Usually, fewer than 1 out of 1,000 of the eggs will become adult walleye.

Predators and Prey

From 1995 to 1997, forty-one gray wolves were brought to Yellowstone National Park, where no wolves had lived for more than 60 years. Since wolves eat elk, scientists collected data to study what happened to the elk population.

When there were no wolves to eat them, the elk population increased greatly. In 1995, there were about 17,000 elk living on the northern range of the park.

By 2004, the number of elk decreased to about 8,000 elk while the number of wolves increased to 171 wolves. By 2013, both elk and wolf populations decreased. By then, there were only about 4,000 elk and 83 wolves.

Scientists wanted the wolf population to grow, but they didn't want the elk population to shrink too much. They are still working to find and keep a healthy balance of wolves and elk in the park.

What other animals in the wild would you like to study using mathematics as a scientist?

Number Uses

Most people use hundreds or thousands of numbers each day. There are numbers on clocks, calendars, car license plates, postage stamps, scales, and so on. These numbers are used in many different ways.

Common ways for using numbers are listed below.

• Numbers are used as **counts.** A count is a number that tells "how many."

Examples

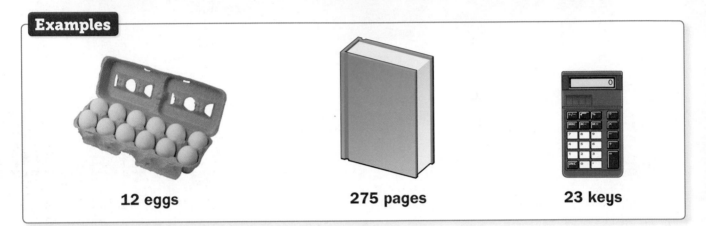

| 12 eggs | 275 pages | 23 keys |

• Numbers are used as **measures.** A measure tells "how much" of something there is. For example, a ruler is used to measure distances.

Examples

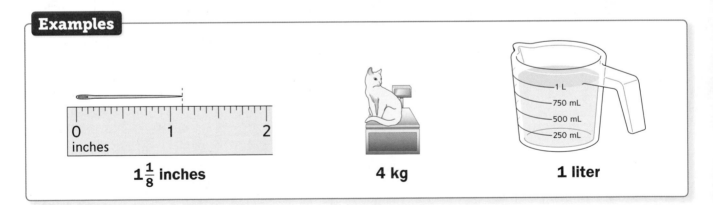

| $1\frac{1}{8}$ inches | 4 kg | 1 liter |

To count whole things, you use the numbers 1, 2, 3, and so on. For measuring, you often need "in-between" numbers. You can use fractions and decimals to name amounts between whole numbers. For example, in the ruler shown above, marks between the inch marks show fractions of inches.

McGraw-Hill Education/Mark Steinmetz

- Numbers are used to show **locations** compared to some starting point.

Examples

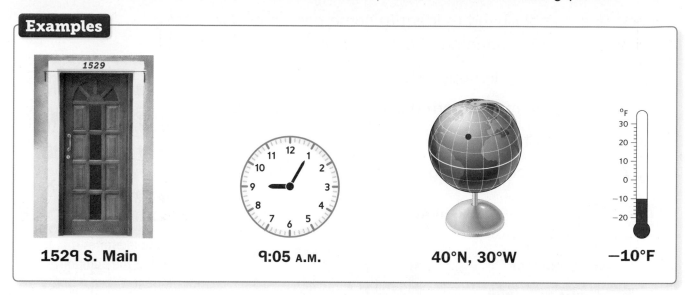

1529 S. Main　　　**9:05 A.M.**　　　**40°N, 30°W**　　　**−10°F**

The street address 1529 S. Main gives a location on Main Street. Clock times are locations in time starting at midnight or noon. The pair of numbers 40°N, 30°W gives a location on Earth's surface. Temperatures give a location on a thermometer starting at 0 degrees. Negative numbers show temperatures that are below 0 degrees. A temperature of −10°F is read as "10 degrees below zero." The numbers −1, −2, −3, and $-\frac{1}{2}$ are all negative numbers.

- Numbers are used to make **comparisons** between counts and measures.

Examples

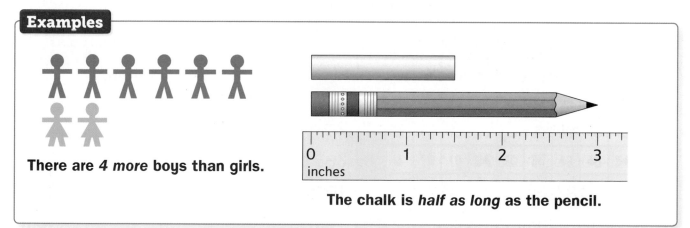

There are *4 more* boys than girls.

The chalk is *half as long* as the pencil.

- Numbers are used as **codes.**

A code is a number used to identify some person or some thing.
Codes are used in phone numbers, credit cards, and ZIP codes.

For example, in the ZIP code **60637:**

> **6** refers to the midwestern part of the United States.

> **06** refers to Chicago.

> **37** refers to a certain neighborhood in Chicago.

Examples

772-555-1212
phone number

license plate number

bar code and ISBN code

Did You Know?

A bar code identifies a product and the company that made it. When a bar code is scanned at a store, the bar code number is sent to the store's computer. The computer looks up the price for that bar code number, and the price is shown on the cash register.

ISBN codes are used to identify books.

Reminder: Most numbers come with a unit or symbol that tells what the number means: 10 cats, 10 inches, 10 A.M., and 10°F mean different things. The unit or symbol shows which meaning you want.

Check Your Understanding

Decide whether each number is used to count, to measure, to show a location, to compare two quantities, or to serve as a code.

1. 25 meters
2. 1-800-555-1212
3. 11/18/15
4. 7 more than
5. 2 kg
6. 237 Church St.
7. $2\frac{1}{4}$ inches
8. 303 children
9. 10 less than
10. 12 noon
11. 7:40 P.M.
12. 0 pink elephants

Check your answers in the Answer Key.

Iconotec/Glow Images

Number Grids

A monthly calendar is an example of a **number grid.** The numbers of the days of the month are listed in order.

The numbers are printed in boxes. The boxes are printed in rows. There are 7 boxes in each row because there are 7 days in a week.

May						
Sun	Mon	Tue	Wed	Thu	Fri	Sat
				1	2	3
4	5	6	7	8	9	10
11	12	13	14	15	16	17
18	19	20	21	22	23	24
25	26	27	28	29	30	31

Part of a number grid like the one you use in class is shown to the right. The numbers are listed in order and printed in rows of boxes. There are 10 boxes in each row. The number in the last box of each row has 0 as the final digit.

Counting forward on a number grid is like reading a calendar. When you reach the end of a line, you go to the next line below and start at the left.

−9	−8	−7	−6	−5	−4	−3	−2	−1	0
1	2	3	4	5	6	7	8	9	10
11	12	13	14	15	16	17	18	19	20
21	22	23	24	25	26	27	28	29	30
31	32	33	34	35	36	37	38	39	40
41	42	43	44	45	46	47	48	49	50
51	52	53	54	55	56	57	58	59	60
61	62	63	64	65	66	67	68	69	70
71	72	73	74	75	76	77	78	79	80
81	82	83	84	85	86	87	88	89	90
91	92	93	94	95	96	97	98	99	100
101	102	103	104	105	106	107	108	109	110

The numbers on a number grid have some simple patterns. These patterns make the grid easy to use.

- When you move *right*, numbers *increase by 1.*

  For example, 15 is 1 more than 14.

- When you move *left*, numbers *decrease by 1.*

  For example, 23 is 1 less than 24.

- When you move *down*, numbers *increase by 10.*

  For example, 37 is 10 more than 27.

- When you move *up*, numbers *decrease by 10.*

  For example, 91 is 10 less than 101.

A number grid, like a number line, can be extended to include all numbers.

Example

This is part of a number grid that shows larger numbers.

511	512	513	514	515	516	517	518	519	520
521	522	523	524	525	526	527	528	529	530
531	532	533	534	535	536	537	538	539	540
541	542	543	544	545	546	547	548	549	550
551	552	553	554	555	556	557	558	559	560
561	562	563	564	565	566	567	568	569	570
571	572	573	574	575	576	577	578	579	580
581	582	583	584	585	586	587	588	589	590

Example

Part of a number grid is shown below.

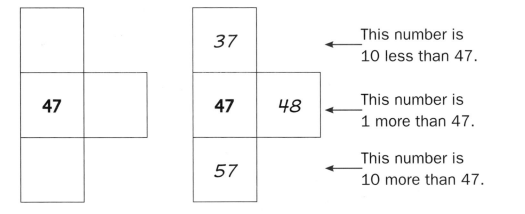

This number is 10 less than 47.

This number is 1 more than 47.

This number is 10 more than 47.

You can use number-grid patterns to fill in the missing numbers.

A number grid can help you add numbers.

Example

48 + 25 = **?**

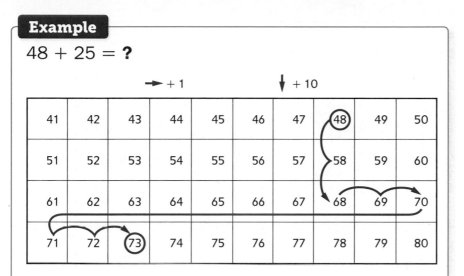

- Start at 48.

- Add 20, or 2 tens, by moving down 2 rows to 68.

- Add 5, or 5 ones, by counting on 5 spaces.

48 + 25 = **73**

Note At the end of a row, go to the beginning of the next row and keep counting.

A number grid can help you find the difference between two numbers.

Example

Find the difference between 37 and 64.

31	32	33	34	35	36	37	38	39	40
41	42	43	44	45	46	47	48	49	50
51	52	53	54	55	56	57	58	59	60
61	62	63	64	65	66	67	68	69	70

- Start at 37.

- Count the number of *tens* going down to 57. There are 2 tens, or 20.

- Count the number of *ones* going right from 57 to 64. Count 1 at 58, 2 at 59, 3 at 60, 4 at 61, and so on. There are 7 ones, or 7.

- The difference between 37 and 64 is 2 tens and 7 ones, or 27.

Number and Operations in Base Ten

Number grids can be used to explore number patterns.

Example

Start with 0. Count by 2s until you reach 100.

									0
1	2	3	4	5	6	7	8	9	10
11	12	13	14	15	16	17	18	19	20
21	22	23	24	25	26	27	28	29	30
31	32	33	34	35	36	37	38	39	40
41	42	43	44	45	46	47	48	49	50
51	52	53	54	55	56	57	58	59	60
61	62	63	64	65	66	67	68	69	70
71	72	73	74	75	76	77	78	79	80
81	82	83	84	85	86	87	88	89	90
91	92	93	94	95	96	97	98	99	100

The blue boxes contain *even* numbers.

The orange boxes contain *odd* numbers.

Check Your Understanding

1. Use the number grid above to find the difference.

   a. Between 16 and 46   **b.** Between 73 and 98

   c. Between 37 and 72

2. Copy the parts of the number grids shown.

   Use number-grid patterns to find the missing numbers.

   a.   **b.**   **c.**

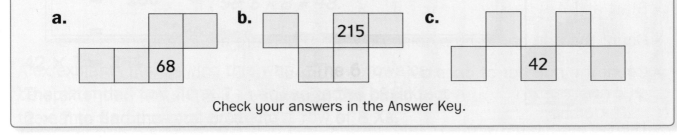

Check your answers in the Answer Key.

Number Lines

A number line is a line with numbers marked in order from left to right. An example is shown below.

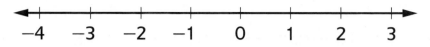

The number 0 is called the **zero point.** All of the spaces between marks are the same length.

The numbers to the right of 0 are called **positive numbers.**
The numbers to the left of 0 are called **negative numbers.**
For example, −3 is called "negative 3." Zero is neither positive nor negative.

Example

The number line below shows the numbers −2, −1, 0, 1, 2, and 3.

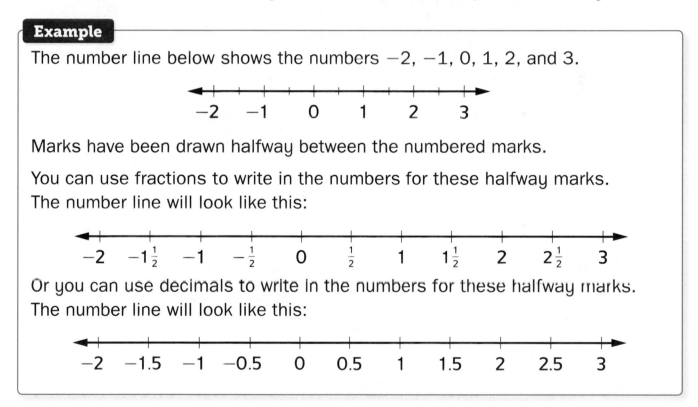

Marks have been drawn halfway between the numbered marks.

You can use fractions to write in the numbers for these halfway marks. The number line will look like this:

Or you can use decimals to write in the numbers for these halfway marks. The number line will look like this:

Every ruler is a number line. If the zero mark is at the end of the ruler, the number 0 may not be printed on the ruler.

On rulers, inches are usually divided into halves, quarters, eighths, and sixteenths. The marks showing fractions of an inch are usually different lengths.

Example

Every thermometer is a number line.

The zero mark on a Celsius scale (0°C) is the temperature at which water freezes.

Negative numbers are shown on the thermometer. A temperature of −16°C is read as "16 degrees below zero."

The marks on a thermometer are evenly spaced. The space between marks is often 2 degrees.

Number lines can also show counting by larger numbers.

Example

This number line shows numbers counting by 10s.

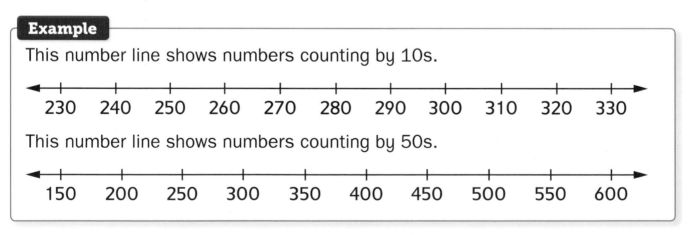

This number line shows numbers counting by 50s.

Comparing Numbers

When two numbers are compared, two results are possible:

- The numbers are **equal.** One number is neither more nor less. Both numbers name the same amount.

- The numbers are **not equal.** One of the numbers is larger than the other.

Different symbols are used to show that the numbers are equal or not equal.

- Use an *equal sign* (=) to show that the numbers are *equal.*

- Use a *greater-than symbol* (>) or a *less-than symbol* (<) to show that the numbers are *not equal.*

Here is one way to remember the meaning of the > and < symbols. Think of each symbol as a mouth. The mouth must be open to swallow the larger number.

$5 > 2 \quad 2 < 5$

Example

The table below lists other examples. Some examples compare numbers, and others compare amounts.

Symbol	Meaning	Examples
=	"equals" "is the same as"	$20 = 4 \times 5$ $12 \text{ inches} = 1 \text{ foot}$ $365 = 300 + 60 + 5$ $5 + 5 = 12 - 2$ $10 = 10$
>	"is greater than"	$7\frac{1}{2} \text{ in.} > 7 \text{ in.}$ $596 > 589$ $1{,}000 > 8 \text{ hundred}$ $714 \text{ mL} > 471 \text{ mL}$
<	"is less than"	$199 < 200$ $506 < 500 + 30 + 6$ $\frac{1}{2} < 1$ $25 \text{ g} < 28 \text{ g}$

Name-Collection Boxes

Any number can be written in many different ways. Different names for the same number are called **equivalent** names.

A **name-collection box** is a diagram used to write names for the same number.

• The name on a label gives a number.

• The names written inside the box are equivalent names for the name on the label.

Example

A name-collection box for 8 is shown to the right. It is called an "8-box."

8		
	2 × 4	~~HHT~~ ///
eight	8 − 0	8 ÷ 1
ocho	2 + 2 + 2 + 2	
	$\frac{8}{1}$	

To form equivalent names for numbers, you can:

• add, subtract, multiply, or divide

• use tally marks or arrays

• write words in English or other languages

Example

A name-collection box for 57 is shown to the right. It is called a "57-box."

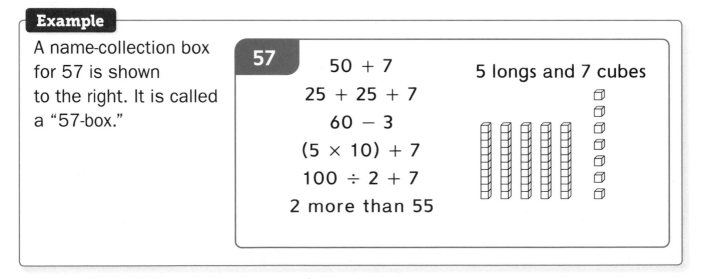

57	
50 + 7	5 longs and 7 cubes
25 + 25 + 7	
60 − 3	
(5 × 10) + 7	
100 ÷ 2 + 7	
2 more than 55	

Each name in the 57-box on the previous page is a different way to say the number 57. This means that we can use an equal sign (=) to write each statement below.

5 longs and 7 cubes

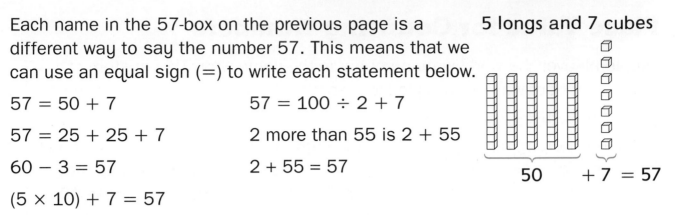

$57 = 50 + 7$

$57 = 25 + 25 + 7$

$60 - 3 = 57$

$(5 \times 10) + 7 = 57$

$57 = 100 \div 2 + 7$

2 more than 55 is $2 + 55$

$2 + 55 = 57$

$50 \qquad + 7 = 57$

Check Your Understanding

1. What name belongs on the label for this name-collection box?

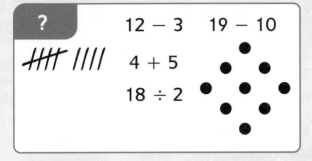

2. Draw a 6-box like the one shown. Write five equivalent names for 6 in your 6-box.

3. List the names that do not belong in this name-collection box. Explain why one of the names you listed doesn't belong.

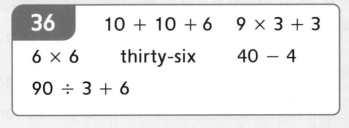

Check your answers in the Answer Key.

Place Value for Counting Numbers

People all over the world write numbers in the same way. This system of writing numbers was invented in India about 1,500 years ago. It is called a **place-value** system.

You can write any number using the ten **digits** 0, 1, 2, 3, 4, 5, 6, 7, 8, and 9. The **place** for a digit, or its position in the number, is very important. The **value** of a digit is how much it is worth. This number system is a **base-10** system because each place in a number is based on 10.

Example

The numbers 72 and 27 use the same digits, a 7 and a 2. But 72 and 27 are different numbers because the 7 and the 2 are in different places.

tens place	**ones place**		**tens place**	**ones place**
7	2		2	7

The digit 2 in 72 is worth 2 (2 ones).

The digit 7 in 72 is worth 70 (7 tens).

The digit 2 in 27 is worth 20 (2 tens).

The digit 7 in 27 is worth 7 (7 ones).

Example

The number 55 uses the digit 5 twice. But the two 5s are in different places.

The 5 in the tens place is worth 50 (5 tens).

The 5 in the ones place is worth 5 (5 ones).

tens place	**ones place**
5	5

Example

Although its value is zero, or nothing, the digit 0 is very important when it is in a place in a number.

tens place	**ones place**		**tens place**	**ones place**
7	0			7

The digit 7 is worth 70 in this number.

With the 0 removed, the digit 7 is only worth 7.

You can write a number in **expanded form** by writing it as the sum of the values of each digit.

Examples

Write each number in expanded form.

72 = 70 + 2   **55** = 50 + 5

27 = 20 + 7   **369** = 300 + 60 + 9

The value of the digit 0 in any place is zero.
There are two ways to show a number with 0 tens in expanded form.

408 = 400 + 0 + 8   or   **408** = 400 + 8

You can use a **place-value chart** to show how much each digit in a number is worth.

Example

The number 3,628 is shown in the place-value chart below.

1,000s	100s	10s	1s
thousands place	hundreds place	tens place	ones place
3	6	2	8

The digit 8 in the 1s place is worth 8 ones, or 8 × 1 = 8.
The digit 2 in the 10s place is worth 2 tens, or 2 × 10 = 20.
The digit 6 in the 100s place is worth 6 hundreds, or 6 × 100 = 600.
The digit 3 in the 1,000s place is worth 3 thousands, or 3 × 1,000 = 3,000.

You can write 3,628 in expanded form by writing it as a sum of the values of each digit.

3,628 = 3,000 + 600 + 20 + 8

3,628 is read as "three thousand, six hundred twenty-eight."

You can use a place-value chart to compare two numbers.

Example

Compare the numbers 6,429 and 6,942. Which number is greater?

1,000s thousands	100s hundreds	10s tens	1s ones
6	4	2	9
6	9	4	2

Start at the left side.
The 1,000s digits *are* the same. They are both worth 6,000.
The 100s digits *are not* the same. The 9 is worth 900, so it is worth more than the 4 that is worth 400.

So 6,942 is the larger number. 6,942 > 6,429

Example

You can use place-value charts to represent the number 35 in different ways.

Amit

10s tens	1s ones
3	5

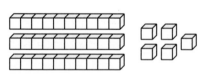

Amit's chart shows 3 tens and 5 ones. This is 30 + 5 = 35.

Meg

10s tens	1s ones
2	15

Meg's chart shows 2 tens and 15 ones. This is 20 + 15 = 35.

Amit and Meg are both correct.

Estimation

An **estimate** is an answer that should be close to an exact answer. You make estimates every day.

- You estimate how long it will take to drive from one place to another.

- You estimate how much money you will need to buy things at the store.

- You estimate how many inches you will grow in the next year.

It may be impossible to find an exact answer. When this happens, you *must* estimate the answer.

Describe some situations when you might estimate.

Example

Police officers and other officials often estimate the number of people at a large event such as a parade, music festival, or street fair. It is not possible to find the exact number of people at the event, so they often use words such as *expect* and *about*.

"Organizers expect more than 50,000 people at the street fair."

Some estimates are called **ballpark estimates.** A ballpark estimate is an answer that may not be very close to the exact answer, but is close enough to help you solve a problem.

Check Your Understanding

1. Estimate how long it takes you to get to school each day. Is it the same everyday?

2. Estimate how long it will take you to do your homework tonight.

3. Estimate how far you can kick or throw a ball. Then measure the actual distance. Compare your estimate to the actual distance.

Estimate When You Don't Need an Exact Answer

An estimate may help you answer a question so that you do not need to find an exact answer.

Example

Carlie needs 500 points to get to the next level of a video game. In the first round she earned 139 points. In the second round she earned 289 points. Can she start the next level of the game?

Carlie can estimate. She can use numbers that are close to the exact amounts, but are easier to add.

	Exact numbers	Close-but-easier numbers
289 is almost 300.	289 points	300 points
139 is almost 150.	139 points	+ 150 points
		450 points

Carlie knows she does not have enough points to make it to the next level.

Example

Ming read 13 pages in 30 minutes. Estimate how many minutes it will take him to read 38 pages.

Use numbers that are close to the exact numbers, but are easier to add or multiply.

	Exact numbers	Close-but-easier numbers
13 is close to 10.	13 pages	10 pages
38 is close to 40.	38 pages	40 pages

Reading 40 pages should take about 4 times as many minutes as reading 10 pages.
4×30 minutes = 120 minutes, or 2 hours

It will take Ming about 2 hours to read 38 pages.

Check Your Understanding

The chess team has $100 to spend on snacks. They want to spend $49 on drinks, $24 on fruit, and $23 on bars. Do they have enough money?

Check your answer in the Answer Key.

Estimate to Check Calculations

Sometimes you want to find an exact answer. Making an estimate helps you check your answer. If your exact answer is not close to your estimate, you should calculate again. One way to estimate is to use **close-but-easier numbers** for the calculations.

Example

Tanesha took a trip. On the first three days she traveled 316 miles, 447 miles, and 489 miles. Tanesha added the three numbers and got 975 miles.

To check that her answer made sense, Tanesha used close-but-easier numbers to estimate the total distance.

	Exact numbers	Close-but-easier numbers
316 is close to 300 miles.	316 miles	300 miles
447 is close to 450 miles.	447 miles	450 miles
489 is close to 500 miles.	489 miles	+ 500 miles
Add the 3 close-but-easier numbers.		1,250 miles

Tanesha's estimate was *not* close to her answer of 975 miles. So she added the three numbers again. This time she got 1,252 miles.

Tanesha's new answer makes more sense. Since it is close to her estimate of 1,250, she probably traveled 1,252 miles, not 975 miles.

Example

James said the difference between 702 and 648 is 54. Is his answer reasonable?

Estimate: 702 is close to 700, and 648 is close to 650.

$700 - 650 = 50$

54 is close to 50, so James's answer is reasonable.

Rounding

Using a number line to **round** numbers is a way to find **close-but-easier numbers.** Here are four steps you can use to round numbers.

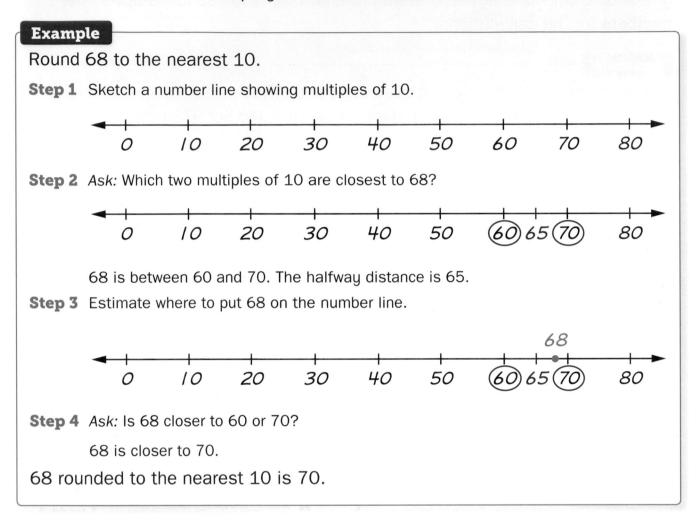

Example

Round 68 to the nearest 10.

Step 1 Sketch a number line showing multiples of 10.

0   10   20   30   40   50   60   70   80

Step 2 *Ask:* Which two multiples of 10 are closest to 68?

0   10   20   30   40   50   ⑥⓪ 65 ⑦⓪   80

68 is between 60 and 70. The halfway distance is 65.

Step 3 Estimate where to put 68 on the number line.

68

0   10   20   30   40   50   ⑥⓪ 65 ⑦⓪   80

Step 4 *Ask:* Is 68 closer to 60 or 70?

68 is closer to 70.

68 rounded to the nearest 10 is 70.

You can use the same four steps shown on page 104 to round to the nearest hundred.

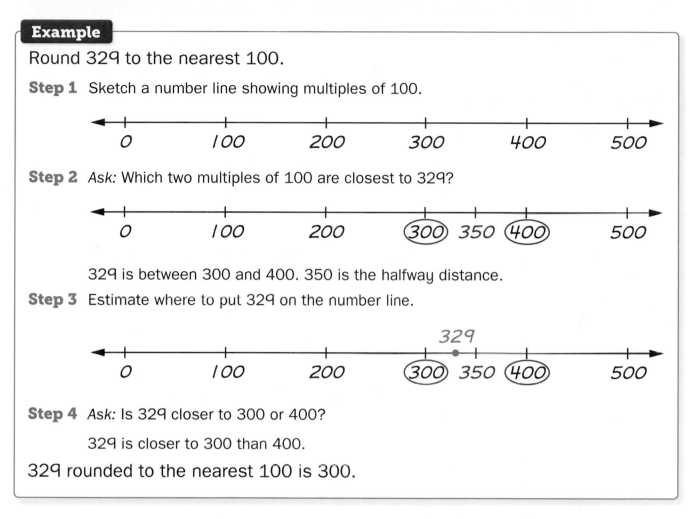

Example

Round 329 to the nearest 100.

Step 1 Sketch a number line showing multiples of 100.

Step 2 *Ask:* Which two multiples of 100 are closest to 329?

329 is between 300 and 400. 350 is the halfway distance.

Step 3 Estimate where to put 329 on the number line.

Step 4 *Ask:* Is 329 closer to 300 or 400?

329 is closer to 300 than 400.

329 rounded to the nearest 100 is 300.

Note Sometimes the number you are rounding is a halfway number. For example, 45 is halfway between 40 and 50 because it is the same distance from both. When this happens, you can decide which way to round based on the situation in the problem. When it isn't clear from the problem which way to round, many people round up. In this case to round to the nearest ten, they would round 45 up to 50.

Rounding to Estimate

Rounding in different ways can result in different reasonable estimates.

Example

Mia had 254 coins and bought 125 more for her collection. About how many coins does Mia have?

$254 + 125 = ?$

One way: Round to the nearest 10.

254 rounds to 250.

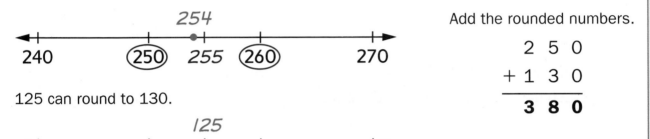

Add the rounded numbers.

```
    2 5 0
 +  1 3 0
 -------
    3 8 0
```

125 can round to 130.

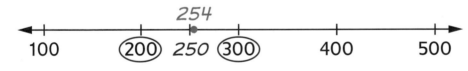

Mia has about **380** coins in her collection.

Another way: Round to the nearest 100.

254 rounds to 300.

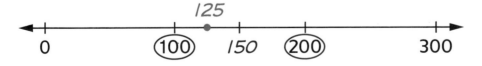

```
    3 0 0
 +  1 0 0
 -------
    4 0 0
```

125 rounds to 100.

Mia has about **400** coins in her collection.

The exact sum is: $254 + 125 = $ **379.** Mia has exactly 379 coins.

Rounding to the nearest 10 and rounding to the nearest 100 both give reasonable estimates for the number of coins she has in her collection.

The example on page 106 shows two different ways to round numbers to estimate an answer to a problem. Deciding whether to round numbers to the nearest 10 or to the nearest 100 depends on how **precise,** or close to the exact answer, you want your answer to be. Rounding numbers to the nearest 10 will generally get an estimate closer to the exact answer than rounding to the nearest 100. Rounding to the nearest 100 may make it easier to estimate using mental math. Both can give reasonable estimates.

Note When rounding halfway numbers, it is often useful to think about the problem situation. In the problem about Mia's coin collection on page 106, think about rounding to the nearest ten. Since 254 was rounded down to 250, rounding 125 up to 130 makes sense to get a closer estimate for the total number of coins.

Check Your Understanding

1. The librarian knows there are 212 books on one shelf and 163 books on another shelf. About how many books are on both shelves?

   a. Estimate the answer by rounding the numbers to the nearest 10.

   b. Estimate the answer by rounding the numbers to the nearest 100.

   c. Find the exact number of books.

   d. Are both estimates reasonable? Which estimate do you prefer? Why?

   Check your answers in the Answer Key.

Basic Facts for Addition and Subtraction

Reading is easier when you know the words by sight. In mathematics, solving problems is easier when you know the basic number facts.

Example

Some basic addition facts:

$6 + 4 = 10$ $0 + 7 = 7$ $3 + 5 = 8$ $9 + 9 = 18$

In an addition problem, the numbers you are adding are called **addends,** and the answer is called the **sum.**

$$\mathbf{7 + 6 = 13}$$
$$\uparrow \quad \uparrow \quad \uparrow$$
$$\text{addends} \quad \text{sum}$$

Some basic subtraction facts:

$10 - 6 = 4$ $7 - 7 = 0$ $8 - 5 = 3$ $18 - 9 = 9$

In a subtraction problem, the result of subtracting one number from another is called the **difference.**

$$\mathbf{13 - 7 = 6}$$
$$\uparrow$$
$$\text{difference}$$

The **facts table** shown to the right is a chart with rows and columns. It shows all of the basic addition and subtraction facts.

Addition/Subtraction Facts Table

+,−	0	1	2	3	4	5	6	7	8	9
0	0	1	2	3	4	5	6	7	8	9
1	1	2	3	4	5	6	7	8	9	10
2	2	3	4	5	6	7	8	9	10	11
3	3	4	5	6	7	8	9	10	11	12
4	4	5	6	7	8	9	10	11	12	13
5	5	6	7	8	9	10	11	12	13	14
6	6	7	8	9	10	11	12	13	14	15
7	7	8	9	10	11	12	13	14	15	16
8	8	9	10	11	12	13	14	15	16	17
9	9	10	11	12	13	14	15	16	17	18

Example

Which addition facts and subtraction facts can you find using the 4-row and the 6-column?

Go across the 4-row while you go down the 6-column. This row and column meet at a square that shows the number 10.

You can use the numbers 4, 6, and 10 to write two addition facts and two subtraction facts:

$4 + 6 = 10$ $\qquad$ $10 - 4 = 6$

$6 + 4 = 10$ $\qquad$ $10 - 6 = 4$

These four facts form a **fact family**.

Addition/Subtraction Facts Table

6-column

+,−	0	1	2	3	4	5	6	7	8	9
0	0	1	2	3	4	5	6	7	8	9
1	1	2	3	4	5	6	7	8	9	10
2	2	3	4	5	6	7	8	9	10	11
3	3	4	5	6	7	8	9	10	11	12
4	4	5	6	7	8	9	10	11	12	13
5	5	6	7	8	9	10	11	12	13	14
6	6	7	8	9	10	11	12	13	14	15
7	7	8	9	10	11	12	13	14	15	16
8	8	9	10	11	12	13	14	15	16	17
9	9	10	11	12	13	14	15	16	17	18

4-row →

Addition Strategies

The addition strategies in this section are shown using basic facts. You can use these same strategies to solve multidigit addition and subtraction problems.

Helper facts are facts you know well. They can help you figure out facts that you do not know.

Doubles are addition facts that have the same number for both addends, such as $6 + 6 = 12$ and $5 + 5 = 10$.

You can think about **near doubles** to solve facts that are close to a double.

Example

$8 + 9 = ?$

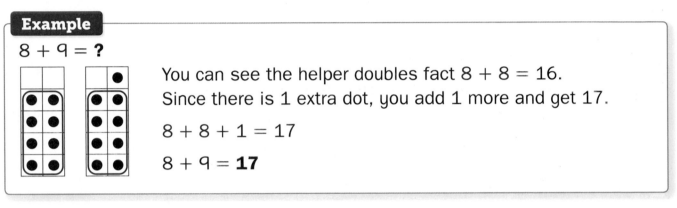

You can see the helper doubles fact $8 + 8 = 16$. Since there is 1 extra dot, you add 1 more and get 17.

$8 + 8 + 1 = 17$

$8 + 9 = \mathbf{17}$

Using near doubles works well when the two addends are close together.

Example

$5 + 7 = ?$ *Think:* $5 + 5 = 10$, so $5 + 5 + 2 = 12$. $5 + 7 = \mathbf{12}$

Combinations of 10 are addition facts with numbers that add to 10, such as $6 + 4 = 10$ and $2 + 8 = 10$. Combinations of 10 can be helper facts.

Sometimes it is easier to break apart one of the addends to make a combination of 10 with the other addend. This is called **making 10.**

Example

$9 + 7 = ?$

You can move 1 dot over to make 10 because $9 + 1 = 10$. There are 6 more dots.

$10 + 6 = 16$

So, $9 + 7 = \mathbf{16}.$

Making 10 works well when one of the addends is close to 10.

Example

8 + 5 = **?**

Think: 8 + 2 = 10. There are 3 more to add. So, 8 + 2 + 3 = 13.

8 + 5 = **13**

The **turn-around rule** says you can add two numbers in either order. Sometimes changing the order makes it easier to solve problems.

Example

4 + 17 = **?**

If you don't know what 4 + 17 is, you can use the turn-around rule to help you, and solve 17 + 4 instead.

17 + 4 is easy to solve by counting on.

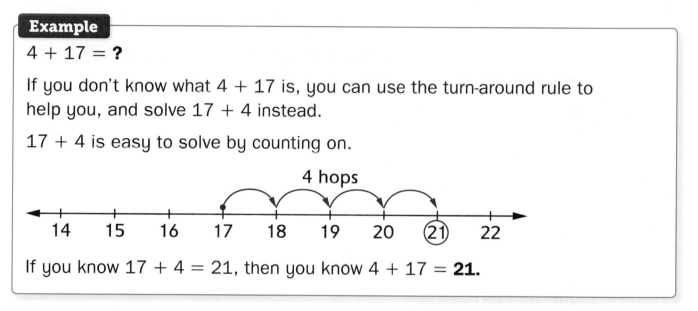

4 hops

If you know 17 + 4 = 21, then you know 4 + 17 = **21.**

Grouping addends together in different ways can make adding easier.

Examples

5 + 8 + 2 = **?**

It can be easier to start with
8 + 2 = 10.

5 + 8 + 2 = **?**
5 + 10 = **15**

2 + 7 + 7 = **?**

It can be easier to start with the double 7 + 7 = 14. Then add 2 more to get 16.

2 + 7 + 7 = **?**
2 + 14 = **16**

Subtraction Strategies

You can **think addition** to subtract. Doubles and combinations of 10 are helpful when you think addition.

Example

15 − 7 = **?**   *Think:* 7 + **?** = 15

You know the double 7 + 7 = 14 is close, so add 1 more to get 7 + **8** = 15. So the difference between 15 and 7 is **8.**

15 − 7 = **8**

In special cases, counting can help you quickly solve a subtraction fact. **Counting back** works well when you are subtracting a small number.

Example

11 − 3 = **?**

Count back 3 from 11.

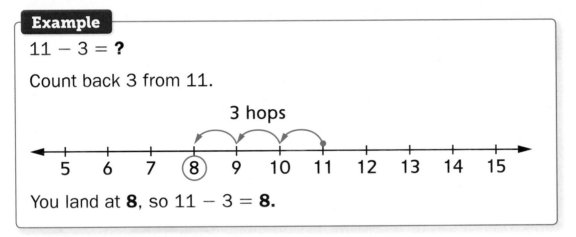

You land at **8**, so 11 − 3 = **8.**

Counting up works well when the numbers are close together.

Example

12 − 9 = **?**

Count up from 9 to 12.

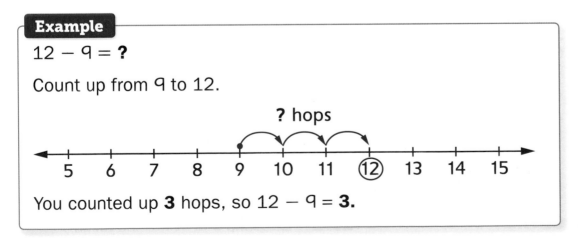

You counted up **3** hops, so 12 − 9 = **3.**

You can **go through 10** to solve many subtraction facts because 10 is a friendly number. You can think of going up or down through 10.

Note A friendly number is a number that is easy for you to work with. Many people use 10 and 5 as friendly numbers.

Example

$17 - 9 = $ **?**

Think: What number should I subtract to get to 10?

Start by subtracting 7 to get down to 10: $17 - 7 = 10$.

You have only subtracted 7. You still need to subtract 2 more to subtract a total of 9: $10 - 2 = $ **8.**

You subtracted a total of 9 and landed at 8. So, $17 - 9 = $ **8.**

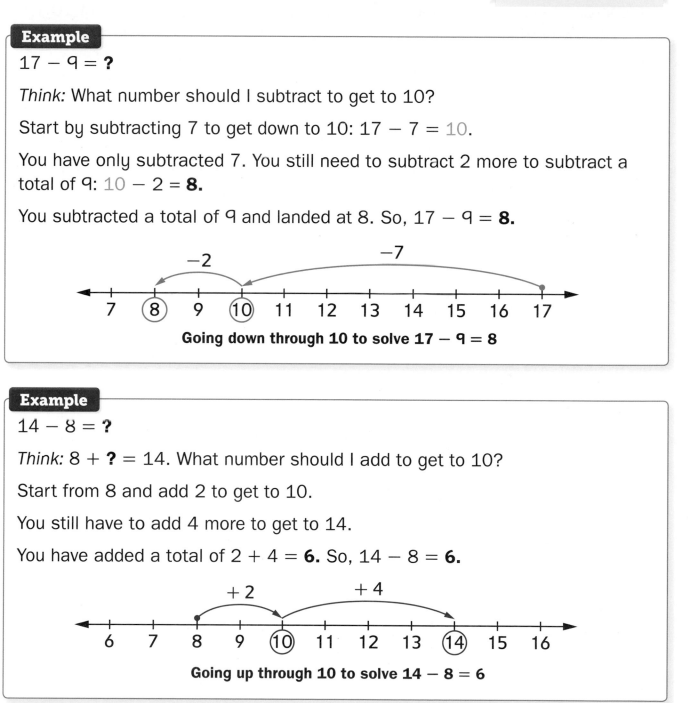

Going down through 10 to solve $17 - 9 = 8$

Example

$14 - 8 = $ **?**

Think: $8 + $ **?** $= 14$. What number should I add to get to 10?

Start from 8 and add 2 to get to 10.

You still have to add 4 more to get to 14.

You have added a total of $2 + 4 = $ **6.** So, $14 - 8 = $ **6.**

Going up through 10 to solve $14 - 8 = 6$

Fact Extensions

Fact extensions are problems with larger numbers that you can solve using basic facts.

Extending Facts to Multiples of 10

You can use basic facts to add or subtract **multiples** of 10. Multiples of 10 and 100 are products of counting numbers and 10 or 100. For example, 30 is a multiple of 10 because $3 \times 10 = 30$ and 600 is a multiple of 100 because $6 \times 100 = 600$.

Example

$50 + 80 = $ **?**

You know $5 + 8 = 13$.

Think: 50 is 5 tens, or 5 [10s]. 80 is 8 tens, or 8 [10s].

$50 + 80 = 5 [10s] + 8 [10s] = 13 [10s]$

13 tens is 130, so $50 + 80 = $ **130.**

Example

$800 - 500 = $ **?**

You know $8 - 5 = 3$.

Think: 800 is 8 hundreds, or 8 [100s]. 500 is 5 hundreds, or 5 [100s].

$8 [100s] - 5 [100s] = 3 [100s]$

So, $800 - 500 = $ **300.**

Example

$150 - 70 = $ **?**

You know $15 - 7 = 8$.

Think: 150 is 15 tens and 70 is 7 tens.

$15 \text{ tens} - 7 \text{ tens} = 8 \text{ tens}$

So, $150 - 70 = $ **80.**

Extending Facts to Higher Decades

You can use basic facts to help solve problems with numbers greater than 10.

Note A *decade* is a multiple of 10 such as 10, 20, 30, and so on.

Examples

5 + 17 = **?**

You know 5 + 7 = 12, so 5 + 17 will be 10 more, or **22.**

5 + 67 = **?**

You know 5 + 7 = 12. There will be a 2 in the ones place, and the tens digit in 67 will *increase* by 1, from 6 tens to 7 tens.

5 + 67 = **72**

36 − 8 = **?**

You know 16 − 8 = 8, so the answer will have 8 in the ones place. The tens digit in 36 will *decrease* by 1, from 3 tens to 2 tens.

36 − 8 = **28**

You can use combinations of 10 to help solve problems with larger numbers.

Examples

32 + **?** = 40

You know 2 + 8 = 10,

so 32 + **8** = 40.

60 − **?** = 54

Think addition: 54 + ? = 60

You know 4 + 6 = 10, so 54 + 6 = 60.

So, 60 − **6** = 54.

Check Your Understanding

Find the sums and differences:

1. 30 + 70 = ?

2. 200 + 700 = ?

3. 100 − 70 = ?

4. 900 − 200 = ?

5. 4 + 28 = ?

6. 430 + ? = 500

Check your answers in the Answer Key.

Partial-Sums Addition

You can use different methods to add. When you use **partial-sums addition,** you can think of each addend in expanded form. Then you can add the 100s, add the 10s, and add the 1s. Finally, add the partial sums you found.

Use an estimate to check whether your answer is reasonable.

> **Note** If you can add the 100s, 10s, and 1s in your head and if you can estimate using mental math, then you don't need to write the steps shown in green.

Example

248 + 187 = **?**

Estimate: 248 is close to 250, and 187 is close to 200.

250 + 200 = 450

The exact sum should be close to 450.

Use partial-sums addition to add:

Think: 248 = 200 + 40 + 8			2	4	8
187 = 100 + 80 + 7		+ 1	8	7
Add the 100s.	200 + 100 →	3	0	0
Add the 10s.	40 + 80 →	1	2	0
Add the 1s.	8 + 7 →		1	5
Add the partial sums.		4	3	5

248 + 187 = **435**

435 is close to the estimate of 450, so 435 makes sense.

Numbers with 4 or more digits can be added in the same way.

You can use base-10 blocks to show how partial-sums addition works.

Example

Use base-10 blocks to add 248 + 187.

Adding the blocks in each column is adding the 100s, 10s, and 1s. Then find the total.

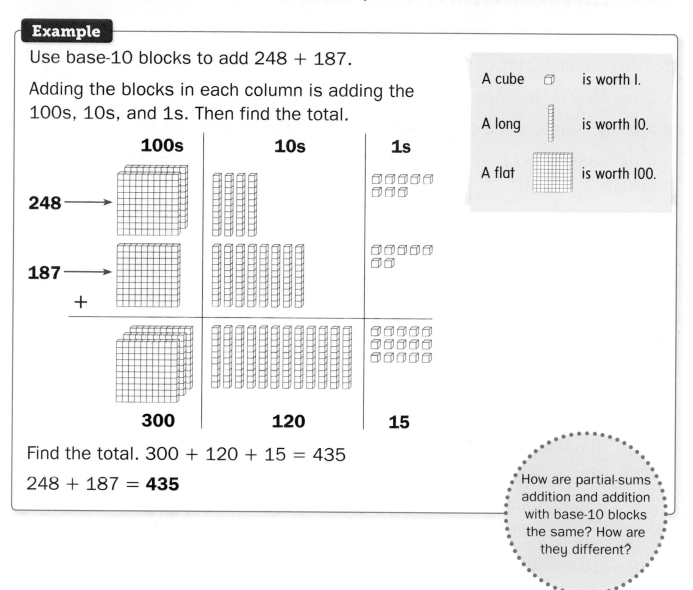

Find the total. 300 + 120 + 15 = 435

248 + 187 = **435**

How are partial-sums addition and addition with base-10 blocks the same? How are they different?

Column Addition

You can use **column addition** to find sums with paper and pencil.

To use column addition:

- Draw lines to separate the 1s, 10s, and 100s places.

- Add each place-value column. Write each sum in its column.

- If the sum of any column is a 2-digit number, make a trade with the column to the left.

You can use an estimate to check whether your answer is reasonable.

Example

$248 + 187 = \,?$

Estimate: 248 is close to 250, and 187 is close to 200.

$250 + 200 = 450$

The exact sum should be close to 450.

Add the numbers in each column.

Trade 10 ones for 1 ten.
Move 1 ten to the tens column.

Trade 10 tens for 1 hundred.
Move the 1 hundred to the hundreds column.

	2	4	8
+	1	8	7
	3	12	15
	3	13	5
	4	3	5

$248 + 187 = \mathbf{435}$

435 is a reasonable answer because it is close to the estimate of 450.

Check Your Understanding

Add. Estimate to check whether your answers are reasonable.

1. $37 + 96$ **2.** $159 + 227$ **3.** $487 + 361$ **4.** $153 + 88$

Check your answers in the Answer Key.

Expand-and-Trade Subtraction

One method you can use to subtract is called **expand-and-trade subtraction.**

> **Example**
>
> $932 - 356 = ?$
>
> Estimate: 932 is close to 950, and 356 is close to 350. $950 - 350 = 600$
>
> | | |
> |---|---|
> | Use expanded form to write the problem so you can see the hundreds, tens, and ones. | $932 \rightarrow 900 + 30 + 2$
 $- 356 \rightarrow 300 + 50 + 6$ |
>
> Look at the hundreds. Since $900 > 300$, you can use this method to subtract.
>
> | | |
> |---|---|
> | Look at the tens. Since $30 < 50$, make a trade.
 Trade 1 hundred for 10 tens. | $\quad\quad\quad 800 \quad 130$
 $932 \rightarrow \cancel{900} + \cancel{30} + 2$
 $- 356 \rightarrow 300 + 50 + 6$ |
>
> *800 + 130 + 2 is still 932 all together. This is a good trade.*
>
> | | |
> |---|---|
> | Look at the ones. Since $2 < 6$, make a trade.
 Trade 1 ten for 10 ones. | $\quad\quad\quad\quad\quad 120$
 $\quad\quad\quad 800 \quad \cancel{130} \quad 12$
 $932 \rightarrow \cancel{900} + \cancel{30} + \cancel{2}$
 $- 356 \rightarrow 300 + 50 + 6$ |
>
> *800 + 120 + 12 is still 932 all together. This is a good trade.*
>
> | | |
> |---|---|
> | Subtract the hundreds, tens, and ones.
 $500 + 70 + 6 = 576$, <br> so 576 is the answer. | $\quad\quad\quad\quad\quad 120$
 $\quad\quad\quad 800 \quad \cancel{130} \quad 12$
 $932 \rightarrow \cancel{900} + \cancel{30} + \cancel{2}$
 $- 356 \rightarrow 300 + 50 + 6$
 $\overline{\quad\quad\quad 500 + 70 + 6 = 576}$ |
>
> $$932 - 356 = \mathbf{576}$$
>
> 576 is close to the estimate of 600, so 576 makes sense.

Check Your Understanding

Use expand-and-trade subtraction to find the difference.
Estimate to check whether your answers are reasonable.

1. $93 - 46$ **2.** $835 - 451$ **3.** $520 - 148$

Check your answers in the Answer Key.

Trade-First Subtraction

Another method you can use to subtract is called **trade-first subtraction.**

To use trade-first subtraction, look at the digits in each place:

- If a digit in the top number is greater than or equal to the digit below it, you do not need to make a trade.

- If any digit in the top number is less than the digit below it, make a trade with the digit to the left.

- After making all necessary trades, subtract in each column.

Use an estimate to decide whether your answer makes sense.

> **Note** If you can keep track of the places in your head, you don't need to draw lines between columns or label the columns. This problem in the example would look like this:
>
> $$
> \begin{array}{rrr}
> & 14 & \\
> 2 & \cancel{15} & 12 \\
> \cancel{3} & \cancel{5} & \cancel{2} \\
> - \quad 1 & 6 & 4 \\
> \hline
> 1 & 8 & 8 \\
> \end{array}
> $$

Example

352 − 164 = **?**

Estimate: 352 is close to 350. You can round 164 to 160.
An estimate is 350 − 160 = 190.

100s	10s	1s
3	5	2
− 1	6	4

Look at the 100s place. Since 300 > 100, there is no trade to make.

100s	10s	1s
	15	
2		
$\cancel{3}$	$\cancel{5}$	2
− 1	6	4

Look at the 10s place. Since 50 < 60, you need to make a trade with the column to the left.

100s	10s	1s
	14	
2	$\cancel{15}$	12
$\cancel{3}$	$\cancel{5}$	$\cancel{2}$
− 1	6	4
1	8	8

Look at the 1s place. Since 2 < 4, you need to make a trade with the column to the left. Now subtract in each column in any order.

352 − 164 = **188**

The answer 188 makes sense because it is close to the estimate of 190.

Base-10 blocks are useful for solving problems. If you don't have blocks, you can draw pictures instead.

Base-10 Blocks Shorthand Pictures		
⬠ = ·	▯ = I	▦ = ☐
cube	long	flat

Example

352 − 164 = **?**

Use pictures of base-10 blocks to model the larger number, 352. Write the number to be subtracted, 164, beneath the block pictures.

Think: Can I remove 1 flat from 3 flats? Yes.

Think: Can I remove 6 longs from 5 longs? No. Trade 1 flat for 10 longs.

Think: Can I remove 4 cubes from 2 cubes? No. Trade 1 long for 10 cubes.

After all of the trading, the blocks look like this:

Now subtract in each column. The remaining blocks show 1 flat, 8 longs, and 8 cubes, which represent 188.

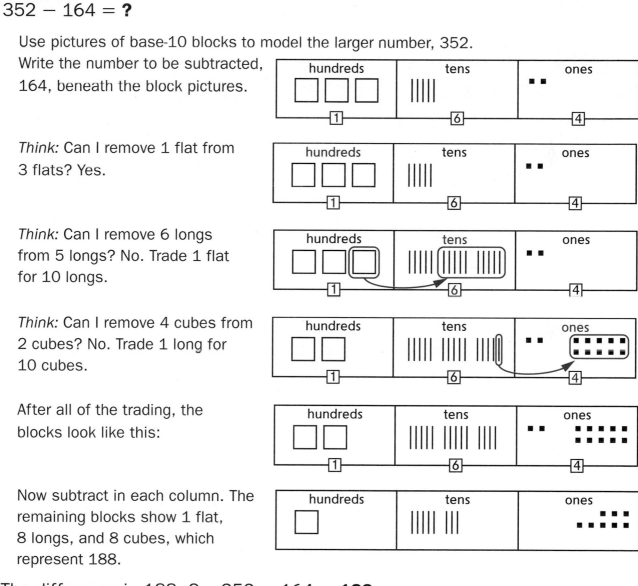

The difference is 188. So, 352 − 164 = **188.**

Counting-Up Subtraction

You can find the difference between two numbers by counting up from the smaller number to the larger number. Subtracting this way is called **counting-up subtraction.** There are many ways to count up. It helps to think of counting to easier numbers, such as numbers that end in zero, or counting by 10s and 100s. One way is to start by counting up to the nearest multiple of 10, then continue counting by 10s and 100s.

You can keep track of your thinking by showing jumps on an open number line.

Example

$325 - 88 = $ **?**

Estimate: $325 - 88$

$\downarrow \quad \downarrow$

$300 - 100 = 200$

The difference should be close to 200.

Draw a line. Mark and label point 88.

88

Think: How can I get from 88 to 325?

Start at 88. Count up 2 to get to 90. Count up 10 to 100. Count up by hundreds to get to 300. Count up 25 more to get to 325.

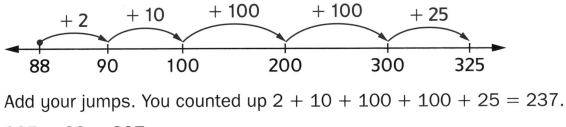

Add your jumps. You counted up $2 + 10 + 100 + 100 + 25 = 237$.

$325 - 88 = $ **237**

The answer 237 makes sense because it is close to the estimate of 200.

Another way to keep track of your thinking as you count up is to write number sentences.

Example

325 − 88 = **?**

Start with the smaller number, 88, and count up to 325. Circle each amount that you count up.

One way:

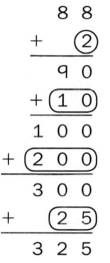

Count up to the nearest 10.

Count up to the nearest 100.

Count up to the largest possible hundred.

Count up to the larger number.

Another way:

88 + ⟮2⟯ = 90

90 + 10 = 100

100 + 200 = 300

300 + 25 = 325

2 + 10 + 200 + 25 = 237

325 − 88 = **237**

Then add the numbers you circled:
2 + 10 + 200 + 25 = 237

You counted up by 237.

325 − 88 = **237**

Check Your Understanding

Use counting-up subtraction to subtract. Estimate to check whether your answers are reasonable.

1. 90 − 33   **2.** 242 − 70   **3.** 742 − 387   **4.** 360 − 179

Check your answers in the Answer Key.

Extended Multiplication Facts

A basic multiplication fact is a **product** of two 1-digit numbers, called **factors.** For example, 8 × 5 = 40 and 3 × 4 = 12 are **basic facts.** When you skip-count by a number, your counts are the **multiples** of that number. The multiples of 10 are 10, 20, 30, 40, and so on.

Examples

Multiples of 10 can be written as a number of groups of ten.

40 = 4 × 10   40 is 4 groups of 10.   40 = 4 [10s]

90 = 9 × 10   90 is 9 groups of 10.   90 = 9 [10s]

Extended multiplication facts have at least one factor that is a multiple of 10. You can use basic facts to solve extended multiplication facts.

Example

3 × 60 = **?**

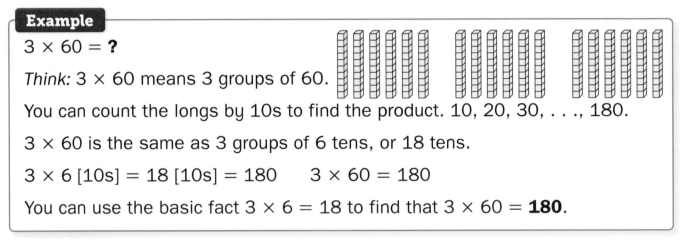

Think: 3 × 60 means 3 groups of 60.

You can count the longs by 10s to find the product. 10, 20, 30, . . ., 180.

3 × 60 is the same as 3 groups of 6 tens, or 18 tens.

3 × 6 [10s] = 18 [10s] = 180   3 × 60 = 180

You can use the basic fact 3 × 6 = 18 to find that 3 × 60 = **180**.

Example

5 × 80 = **?**

Use the basic fact 5 × 8 = 40.   5 × 80 = 5 groups of 8 [10s]

5 × 8 [10s] = 40 [10s] = 400   5 × 80 = **400**

Example

7 × 2 = **?**   7 × 2 = **14**

7 × 20 = **?**   7 × 20 = 7 × 2 [10s] = 14 [10s] = **140**

7 × 200 = **?**   7 × 200 = 7 × 2 [100s] = 14 [100s] = **1,400**

A History of Counting and Calculating

In the thousands of years since people began to count, we have developed many different number systems and tools for counting and calculating.

We learn to count on our fingers at a very young age.

In China, a method has been developed to count up to 100 thousand using one hand and up to 10 billion using two hands. Do you see the pattern?

Many of us use our hands as simple calculators when we start learning to add and subtract. As we learn math facts, we rely on our fingers less and less.

Writing Numbers

Different societies have used different types of systems for writing numbers. The number system that we use is the base-10 system.

The Babylonians of the Middle East used 60 as the base for their number system. We still use base-60 today in measuring time. For example, there are 60 minutes in an hour and 60 seconds in a minute.

This Babylonian clay tablet shows ancient numbering.

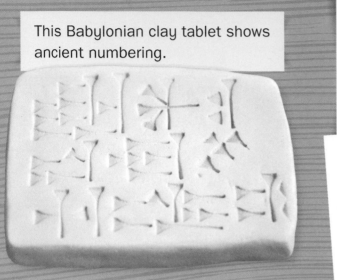

The ancient Maya of Central America chose 20 as the base for their number system, perhaps because they counted using their fingers and toes.

Roman numerals came into use more than 2,000 years ago. They are still used today for formal purposes. For example, people use Roman numerals on buildings and clocks and for sporting events.

The year the Declaration of Independence was signed, 1776, is shown on a dollar bill in Roman numerals at the bottom of the pyramid.

The Western world today uses Arabic numerals. This base-10 system requires ten different digits as shown on the calulator. Arabic numerals, like Mayan numerals, have a symbol for zero. Roman numerals do not.

(tl)©Kenneth Norton/Alamy; (tr)andrewsafonov/iStock/360/Getty Images; (bl)McGraw-Hill Education; (br)Dieter Spannknebel/Stockbyte/Getty Images

Tools to Count and Calculate

The tools people use to count and calculate have improved with advances in technology.

The first tools used for counting were probably fingers. Do you think there is a connection between our ten fingers and the base-10 number system we use today?

Neighborhood Pets

Pets	Tallies
Dogs	＃＃＃
Cats	＃＃＃ //
Rabbits	//

Tally marks, another ancient tool for counting, are still used today. When used to count in base 10, they're written in groups of five, with the fifth tally mark crossing over the first four. Tally marks are a quick and efficient tool to keep track of counts. What is the total number of pets in the neighborhood?

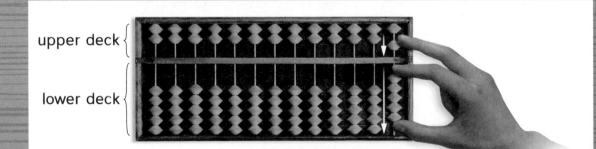

upper deck
lower deck

The bead abacus was invented in China hundreds of years ago. Beads are counted by moving them to the beam that separates the upper and lower decks. After 5 beads are counted in the lower deck, the result is "carried" to the upper deck. Besides counting, the bead abacus can be used for calculations such as adding and subtracting.

Since the abacus, other calculation devices have been invented.

John Napier invented a device, known as "Napier's bones," in the late 1500s. The "bones" were often made of wood or bone, and they were used like multiplication tables.

In the 1600s, William Oughtred created the slide rule. The slide rule is used to perform multiplication and division by sliding the two scales by one another. It was the most commonly used calculation tool in science and engineering until the 1970s.

Mechanical calculators were first developed in the 1600s. Through the 20th century, mechanical calculators were developed that could automatically add, subtract, multiply, and divide.

Electronic Calculators

The abacus and other calculating tools are being replaced by electronic calculators in most parts of the world. Today's electronic calculators can perform more advanced operations than older calculating tools.

Handheld calculators were first sold in 1971. They are smaller and more portable than the abacus. One or more silicon chips act as the "brain" of the calculator.

Today, many handheld calculators can calculate faster than the most powerful computers of 40 years ago.

As time goes on, humans will invent more powerful ways to calculate. But no matter how advanced technology becomes, people will continue to use their fingers and brains for counting and calculating.

What methods do you use to count, add, subtract, multiply, and divide?

filonmar/E+/Getty Images

Mediaphotos/iStock/Getty Images Plus/Getty Images

Fractions

Fractions were invented thousands of years ago because people needed to name numbers that were between whole numbers. Fractions are used to name parts of wholes.

Four children are sharing this veggie pizza equally. Each child will get less than 1 whole pizza. What fraction names the amount of pizza each child will get?

a whole veggie pizza

The whole veggie pizza is divided into 4 equal shares. Each share is called a *fourth* or a *quarter*.

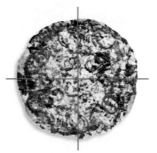

Here are some ways you can describe the fraction of the veggie pizza that each child will get:

1-fourth   one-fourth   1 quarter   1 out of 4 equal shares

Here are some ways you can describe the whole veggie pizza:

4-fourths   four-fourths   4 quarters   4 out of 4 equal shares

Each child would not get the same fraction of the pizza if it were cut into unequal parts.

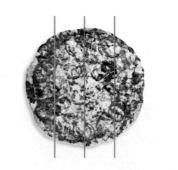

University of Chicago

Example

One part of the garden has flowers. The rest of the garden has vegetables. How much of the garden has vegetables?

the whole garden

the whole garden divided into 3 equal parts

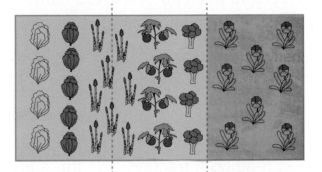

Less than the whole garden has vegetables. A fraction is needed to name the amount of the garden that has vegetables.

When the garden is divided into 3 equal parts, the equal-size parts are named *thirds*.

Here are ways you can describe the 2 equal parts of the garden growing vegetables:

2-thirds   two-thirds   2 out of 3 equal parts

Fractions are needed to describe amounts between whole numbers that are greater than one.

Example

A recipe calls for the juice from one and a half oranges.

That's more than one orange but less than two oranges.

You can describe the number of oranges as one and one-half oranges.

Other ways to name fractions greater than one are described on pages 137–138.

Reading and Writing Fractions

A **fraction** names a part of a whole. You can write a fraction in different ways. For example, three-fourths can be written as:

three of four equal parts   3-fourths $\dfrac{3}{4}$

In fractions such as $\dfrac{3}{4}$, the top number and the bottom number work together to describe the amount of the whole that the fraction represents.

- The **denominator,** the bottom number, describes how many equal parts it takes to make the whole and tells the size of each part.

- The **numerator,** the top number, describes the number of equal-size parts that are being considered.

When reading a fraction, say the numerator first. Then say the size of the equal parts represented by the denominator.

numerator ⟶ $\dfrac{3}{4}$ "three-fourths"
denominator ⟶

Example

Write a fraction to describe the part of this square that is shaded.

The whole is the square.
The square is divided into 8 equal parts.

The size of one part is an eighth. Each part is $\dfrac{1}{8}$ of the square.

Three of the $\dfrac{1}{8}$ parts are shaded: $\dfrac{1}{8} + \dfrac{1}{8} + \dfrac{1}{8} = \dfrac{3}{8}$ (three-eighths).

$\dfrac{3}{8}$ of the square is shaded.

$\dfrac{3}{8}$

3 ⟵ The *numerator* 3 tells the number of *shaded* parts.

8 ⟵ The *denominator* 8 tells the number of equal parts that make the *whole* square.

Example

Write a fraction name for the shaded amount of the large rectangle.

The large rectangle is divided into 6 equal parts, or sixths. Five of the sixths are shaded.

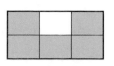

$\dfrac{5}{6}$ of the large rectangle is shaded.

Using Fraction Circles

You can use fraction circles to represent fractions.

The red circle is the largest fraction circle piece, and one light green piece is the smallest fraction circle piece. You can put pieces together to make circles.

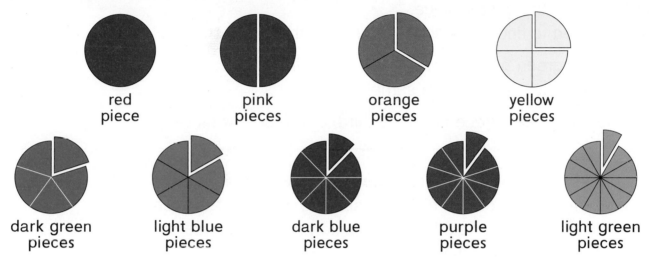

red piece	pink pieces	orange pieces	yellow pieces

dark green pieces	light blue pieces	dark blue pieces	purple pieces	light green pieces

The fraction name for any of the fraction circle pieces depends on which fraction circle piece is the **whole**.

Example

The red fraction circle piece is the whole. Which fraction circle piece is one-third of the whole? Explain how you know.

The whole is the red piece.

Think: $\frac{1}{3}$ is one out of three equal parts. I need to find three equal-size pieces that cover the whole.

Three orange pieces cover the whole red piece. Each orange piece is the same size.

So, one orange piece is $\frac{1}{3}$ (one-third) of the red circle.

Example

The yellow fraction circle piece is the whole. Which fraction piece is $\frac{1}{3}$ (one-third) of the whole? Explain how you know.

The whole is the yellow piece.
Think: I need to find three equal-size pieces that cover the whole.

Three light green pieces cover the whole yellow piece.
Each light green piece is the same size.

So one light green piece is $\frac{1}{3}$ (one-third) of the yellow piece.

The fraction circle piece that represents $\frac{1}{3}$ (one-third) is *not* the same when the wholes are different sizes. One-third of the red piece is *greater than* one-third of the yellow piece because one red piece is *larger* than one yellow piece.

Fraction circle pieces can have different fraction names depending on the size of the whole.

Example

The pink piece is the whole. Show $\frac{1}{2}$ of the whole. Explain how you know.

When the pink piece is the whole, two yellow pieces cover the whole. Each yellow piece is $\frac{1}{2}$ of the pink piece.

The yellow piece can be 1 whole. It can also be 1-half. It can have many other fraction names depending on the size of the whole.

Representing Fractions Greater Than One

Fractions can be used to name numbers greater than one.

You can use fraction circle pieces to represent and name a fraction greater than one. One way is to count unit fractions. A **unit fraction** names one equal part of the whole.

Example

The red piece ● is the whole. What fraction can name the amount shown?

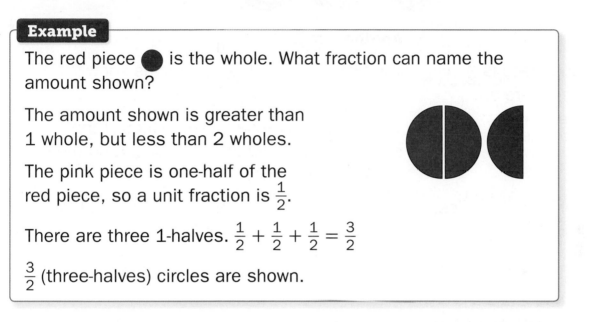

The amount shown is greater than 1 whole, but less than 2 wholes.

The pink piece is one-half of the red piece, so a unit fraction is $\frac{1}{2}$.

There are three 1-halves. $\frac{1}{2} + \frac{1}{2} + \frac{1}{2} = \frac{3}{2}$

$\frac{3}{2}$ (three-halves) circles are shown.

Another way to name the 3 pink pieces is to name the number of wholes first, and then use a fraction to name the remaining parts.

Example

The red piece ● is the whole. What fraction can name the amount shown?

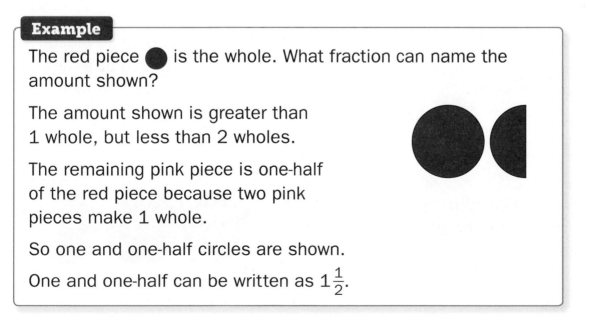

The amount shown is greater than 1 whole, but less than 2 wholes.

The remaining pink piece is one-half of the red piece because two pink pieces make 1 whole.

So one and one-half circles are shown.

One and one-half can be written as $1\frac{1}{2}$.

Example

The red piece 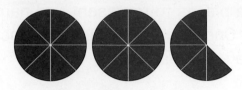 is the whole. What is a fraction name for the amount shown?

One way:

The dark blue piece is $\frac{1}{8}$ of the whole. There are 21 dark blue pieces.

The amount shown is $\frac{21}{8}$ of the whole.

$\frac{21}{8}$ is greater than 2 wholes, but less than 3 wholes.

Another way:

The two complete fraction circles show 2 wholes.

There are 5 more dark blue pieces. They each show one-eighth of the whole, so they make $\frac{5}{8}$ of a whole.

In all, two and five-eighths circles are shown. Two and five-eighths can be written as $2\frac{5}{8}$.

$\frac{21}{8}$ is the same amount as $2\frac{5}{8}$ of the same size whole.

Check Your Understanding

Write a fraction name for each of the amounts shown below.

1. The red fraction circle piece ● is the whole. Name a fraction for the amount shown below.

2. The pink fraction circle piece ◗ is the whole. Name a fraction for the amount shown below.

3. Which is larger? 1-fourth of a red fraction circle piece or 1-fourth of a pink fraction circle piece?

How do you know?

4. The red fraction circle piece ● is the whole. Name a fraction for the amount shown to the right.

Check your answers in the Answer Key.

Using Fraction Strips

A fraction strip represents a **whole.**

1 whole

This fraction strip rectangle is called the *whole, one whole,* or *one.*

These strips can be folded into equal-size pieces to represent fractions. A **unit fraction** names one equal part of the whole.

This fraction strip is partitioned into thirds. The unit fraction is $\frac{1}{3}$.

$\frac{1}{3}$	$\frac{1}{3}$	$\frac{1}{3}$

You can use fraction strips to show other fractions by counting the number of parts shown.

Example

Name a fraction represented by this fraction strip.

The amount shown is greater than 0 but less than one whole fraction strip. Each equal-size piece shows 1-third of a whole fraction strip.

$\frac{1}{3}$	$\frac{1}{3}$

Count: 1-third, 2-thirds. Two-thirds of the whole fraction strip is shown.

$$\frac{1}{3} + \frac{1}{3} = \frac{2}{3}$$

You can use fraction strips to show fractions greater than one whole by combining them with other fraction strips.

Example

Name a fraction represented by these fraction strips.

$\frac{1}{4}$	$\frac{1}{4}$	$\frac{1}{4}$	$\frac{1}{4}$

$\frac{1}{4}$	$\frac{1}{4}$	$\frac{1}{4}$

The amount shown is greater than 1 but less than 2. Count the unit fractions. There are 7-fourths. The fraction is written 7-fourths, or $\frac{7}{4}$.

You can also think about the wholes first, then count on to name the fraction for the remaining parts. In all, 1 whole and 3-fourths strips are shown. You can write one and three-fourths as $1\frac{3}{4}$.

Using Fractions to Name Points on a Number Line

Number lines are used to show distance. The distance from 0 to 1 on a number line is called a **whole.** Equal-size distances within the whole are called parts.

To show a distance traveled on the number line, start at 0, move the given distance, and stop. The stopping point is marked at the end of the distance traveled.

Example

The triangle below traced a path beginning at 0 and ending at 1. The location of the point is 1.

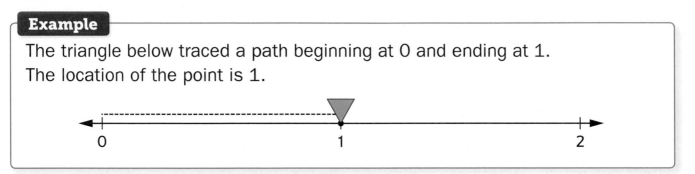

You can use fractions to name points on a number line.

Locating Numbers Less Than One

Number lines can be partitioned, or divided into parts, to show distances from 0 that are less than 1.

Example

What is the location of point A?

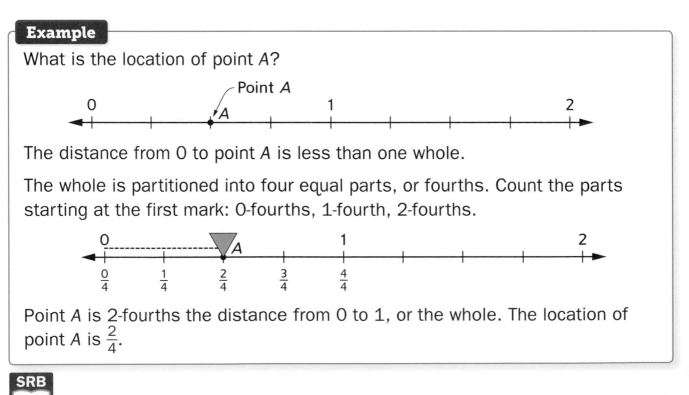

The distance from 0 to point A is less than one whole.

The whole is partitioned into four equal parts, or fourths. Count the parts starting at the first mark: 0-fourths, 1-fourth, 2-fourths.

Point A is 2-fourths the distance from 0 to 1, or the whole. The location of point A is $\frac{2}{4}$.

Example

What fraction can name point *B*?

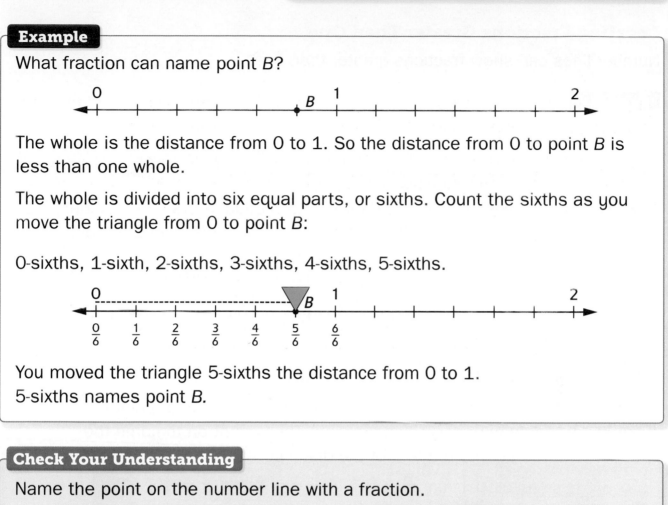

The whole is the distance from 0 to 1. So the distance from 0 to point *B* is less than one whole.

The whole is divided into six equal parts, or sixths. Count the sixths as you move the triangle from 0 to point *B*:

0-sixths, 1-sixth, 2-sixths, 3-sixths, 4-sixths, 5-sixths.

You moved the triangle 5-sixths the distance from 0 to 1.
5-sixths names point *B*.

Check Your Understanding

Name the point on the number line with a fraction.

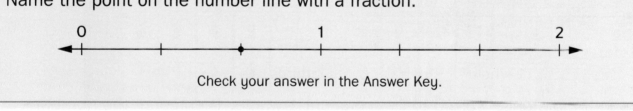

Check your answer in the Answer Key.

Locating Fractions Greater Than One

Number lines can show fractions greater than one.

Example

What is the location of point C?

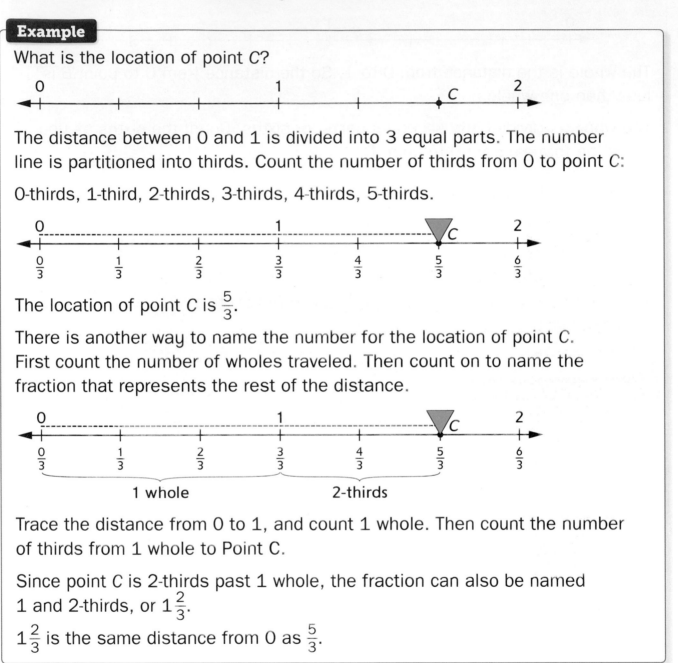

The distance between 0 and 1 is divided into 3 equal parts. The number line is partitioned into thirds. Count the number of thirds from 0 to point C:

0-thirds, 1-third, 2-thirds, 3-thirds, 4-thirds, 5-thirds.

The location of point C is $\frac{5}{3}$.

There is another way to name the number for the location of point C. First count the number of wholes traveled. Then count on to name the fraction that represents the rest of the distance.

Trace the distance from 0 to 1, and count 1 whole. Then count the number of thirds from 1 whole to Point C.

Since point C is 2-thirds past 1 whole, the fraction can also be named 1 and 2-thirds, or $1\frac{2}{3}$.

$1\frac{2}{3}$ is the same distance from 0 as $\frac{5}{3}$.

Example

What fraction can name point *D*?

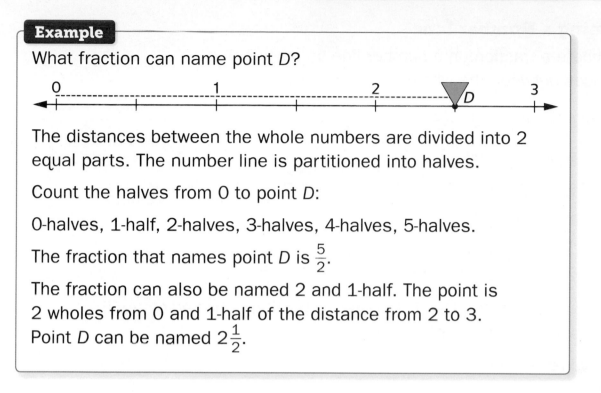

The distances between the whole numbers are divided into 2 equal parts. The number line is partitioned into halves.

Count the halves from 0 to point *D*:

0-halves, 1-half, 2-halves, 3-halves, 4-halves, 5-halves.

The fraction that names point *D* is $\frac{5}{2}$.

The fraction can also be named 2 and 1-half. The point is 2 wholes from 0 and 1-half of the distance from 2 to 3. Point *D* can be named $2\frac{1}{2}$.

Note To determine the denominator of a point on a number line, remember that the whole is the distance between 0 and 1. So you need to count just the equal parts (intervals) between 0 and 1 to find the denominator, rather than than all of the intervals on the number line.

Check Your Understanding

Name the point on the number line with a fraction.

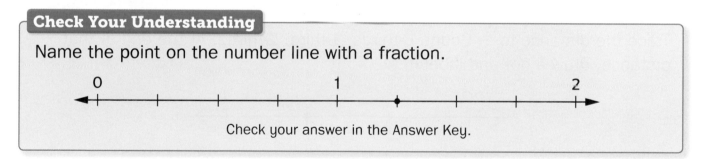

Check your answer in the Answer Key.

Partitioning a Number Line

You can locate a fraction on a number line by partitioning the distances between the wholes.

Example

Show $\frac{2}{3}$ on a number line.

Use a number line from 0 to a whole number larger than the fraction.

$\frac{2}{3}$ is between 0 and 1.

Look at the denominator. Use tick marks to partition each whole on the number line into equal parts.

The denominator shows thirds. Divide the distance between 0 and 1 into 3 equal parts.

Start at 0. Trace the distance of the fraction you want to locate. Make a dot and label the point at the end of the distance.

Trace the distance to $\frac{2}{3}$. Count: 0-thirds, 1-third, 2-thirds. At the end of that distance, draw a dot and label it $\frac{2}{3}$.

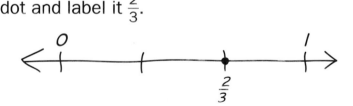

Check Your Understanding

Sketch number lines and show the location of each fraction.

1. $\frac{1}{2}$ **2.** $\frac{3}{4}$ **3.** $\frac{4}{3}$

Check your answers in the Answer Key.

Fraction Number-Line Poster

The Fraction Number-Line Poster can help you partition your own number lines. You can use it to find **equivalent fractions** and to compare fractions.

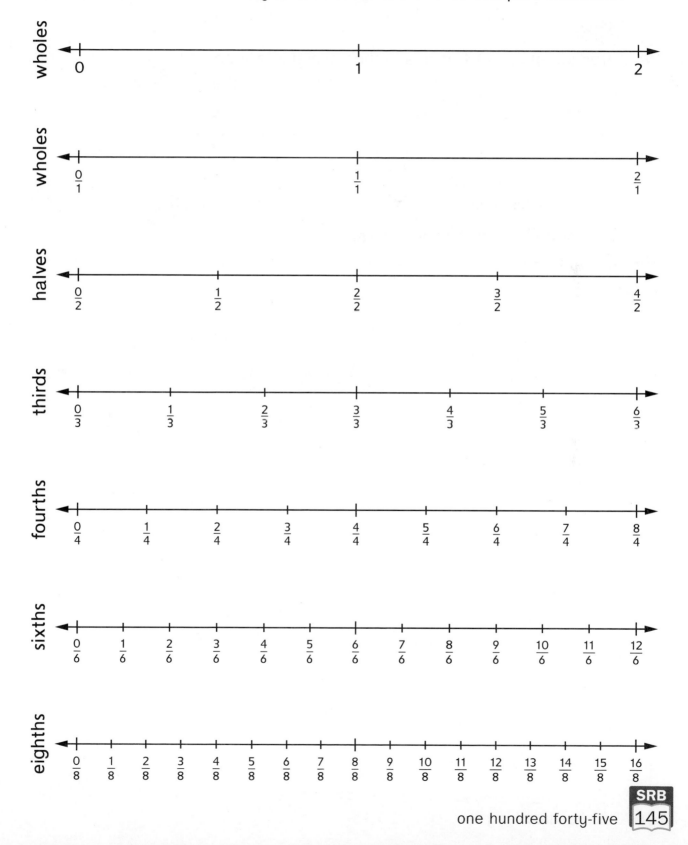

Using Fractions to Name Part of a Collection

Fractions can be used to name part of a whole collection. A collection is a group of items.

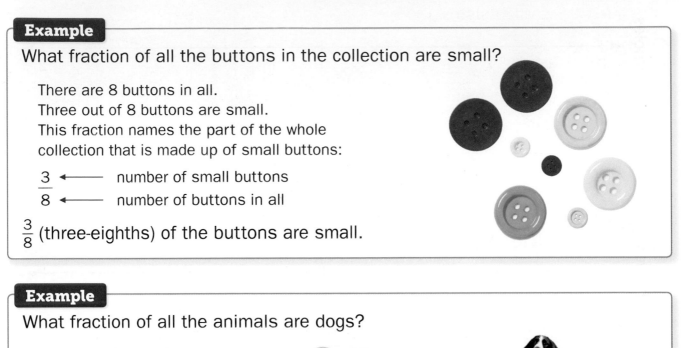

Example

What fraction of all the buttons in the collection are small?

There are 8 buttons in all.
Three out of 8 buttons are small.
This fraction names the part of the whole
collection that is made up of small buttons:

$\dfrac{3}{8}$ ←——— number of small buttons

←——— number of buttons in all

$\dfrac{3}{8}$ (three-eighths) of the buttons are small.

Example

What fraction of all the animals are dogs?

There are 6 animals.
Four animals are dogs.
Four out of 6 animals are dogs.

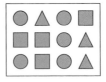

The fraction of animals that
are dogs is $\dfrac{4}{6}$ (four-sixths).

Example

What fraction of the shapes are circles?

There are 12 shapes in all.
Five of the shapes are circles.
Five out of 12 shapes are circles.

The fraction of shapes that are circles
is $\dfrac{5}{12}$ (five-twelfths).

You can solve number stories about collections.

Example

Four sisters collect 8 toy frogs. They share them equally. What fraction of the frogs does each sister get?

Each sister gets $\frac{2}{8}$ of the frogs.

Example

Three dogs are given an equal share of 12 bones. What fraction of the bones does each dog get?

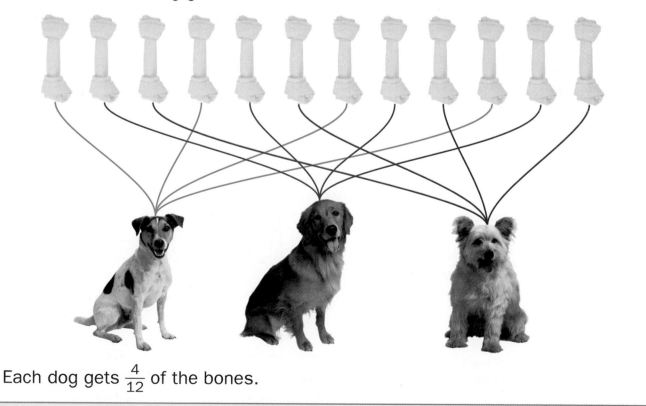

Each dog gets $\frac{4}{12}$ of the bones.

Uses of Fractions

Fractions are useful in everyday life. You can find many fractions in the world around you.

Examples

Signs for Distance	Fractions can be used to tell how far it is to a destination. A scenic view is $\frac{3}{4}$ mile ahead.	**Scenic View** $\frac{3}{4}$ mile ahead
Cooking Recipes	Fractions are often used to tell how much of each ingredient to include in a recipe. Ingredients in this recipe are shown in fractions of cups, tablespoons, and teaspoons.	**Jambalaya** $\frac{3}{4}$ cup rice 4 ounces each of sausage and chicken 4 cups peppers $1\frac{2}{3}$ cups chopped onions $1\frac{1}{2}$ tablespoons chopped thyme $\frac{1}{8}$ teaspoon salt
Measuring Spoons	Fractions are used to measure small amounts of baking or cooking ingredients. This measuring spoon holds $\frac{1}{2}$ teaspoon.	$\frac{1}{2}$ tsp
Rulers	Fractions help describe more precise measures. This key is about $2\frac{1}{4}$ inches long.	0 1 2 inches

Lauri Patterson/E+/Getty Images

You can use fractions to describe the world around you.

Examples

Telling Time | Fractions are often used to describe a part of an hour.

It is **half** past seven.

Money | Fractions can be used to describe amounts of money.

It takes 4 quarters to make a dollar. This coin is $\frac{1}{4}$, or 1-quarter, of a dollar.

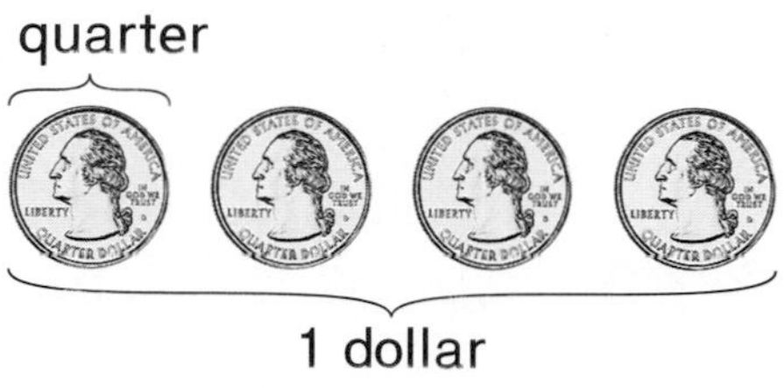

1 quarter

1 dollar

Measurement Descriptions | Fractions can be used to describe amounts of liquids. This liter beaker is filled halfway with water. There is $\frac{1}{2}$ liter of water.

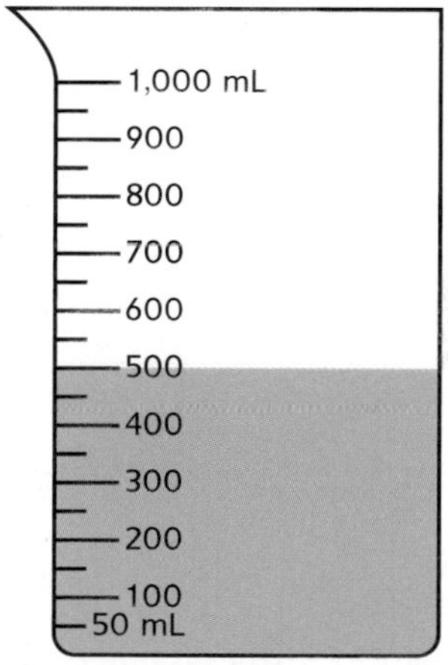

- 1,000 mL
- 900
- 800
- 700
- 600
- 500
- 400
- 300
- 200
- 100
- 50 mL

Comparing Numbers | Fractions can be used to compare amounts.

Alex has $\frac{1}{2}$ as many shells as Faith.

Kay has $\frac{1}{3}$ as many shells as Alex.

Faith:

Alex:

Kay:

Equivalent Fractions

Fractions that name the same amount or name the same distance from 0 are called **equivalent fractions.** Equivalent fractions are equal because they name the same number.

These two circles are the same size.

The first circle is divided into 2 equal parts. The second is divided into 6 equal parts. The top half of each circle is shaded.

The shaded amounts are the same. So 1-half of the circle is the same amount as 3-sixths of the circle.

The fractions $\frac{1}{2}$ and $\frac{3}{6}$ are equivalent fractions. Write $\frac{1}{2} = \frac{3}{6}$.

Example

Eight children go to a party. Two are girls. Six are boys.

$\frac{1}{4}$ of the children are girls.

$\frac{2}{8}$ of the children are girls.

$\frac{1}{4}$ of the children is the same as $\frac{2}{8}$ of the children.

The fractions $\frac{1}{4}$ and $\frac{2}{8}$ are equivalent fractions. Write $\frac{1}{4} = \frac{2}{8}$.

To be equivalent, parts need to be the same size, but not the same shape.

Example

This sandwich can be cut into equal parts in different ways.

The halves are equivalent although they are not the same shape. Each is $\frac{1}{2}$ of the same-size whole. They are the same amount of sandwich.

Example

You can use fraction strips to show equivalent fractions.

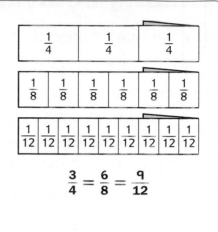

3 parts of the fourths fraction strip is the same size as 6 parts of the eighths fraction strip and 9 parts of the twelfths fraction strip.

The fractions $\frac{3}{4}$, $\frac{6}{8}$, and $\frac{9}{12}$ all name the same amount of the whole.

They are all equivalent fractions.

$$\frac{3}{4} = \frac{6}{8} = \frac{9}{12}$$

Example

The marks on the inch scale of a ruler can be named with equivalent fractions.

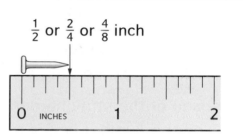

$\frac{1}{2}$ or $\frac{2}{4}$ or $\frac{4}{8}$ inch

The head of the nail is at 0 and the tip of the nail is at $\frac{1}{2}$. The length of this nail is $\frac{1}{2}$ inch. You can determine other names for the length of the nail by looking at the tick marks for smaller intervals: $\frac{2}{4}$ in. or $\frac{4}{8}$ in.

$\frac{1}{2}$, $\frac{2}{4}$, and $\frac{4}{8}$ inch all name the same length on the ruler.

$\frac{1}{2}$, $\frac{2}{4}$, and $\frac{4}{8}$ are equivalent fractions.

Check Your Understanding

Use drawings or fraction tools to decide whether the given pairs of fractions are equivalent.

1. $\frac{1}{2}$ or $\frac{2}{4}$ of a pizza

2. $\frac{3}{4}$ or $\frac{5}{8}$ of a mile

Check your answers in the Answer Key.

Finding Equivalent Fractions

You can use fraction tools to find **equivalent fractions**.

Number lines can show equivalent fractions when the wholes are the same length on each number line. When fractions are equivalent, they are the same distance from 0.

To use a Fraction Number-Line Poster to find equivalent fractions, line up a straightedge vertically next to the point for a fraction.

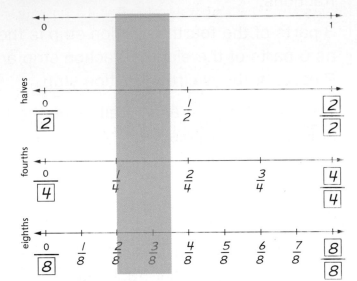

If the edge lines up with the point for another fraction, the fractions are equivalent because they are the same distance from 0.

The side of the straightedge lines up with $\frac{1}{4}$ and $\frac{2}{8}$. So the distance from 0 to $\frac{1}{4}$ on the number line is the same as the distance from 0 to $\frac{2}{8}$.

$\frac{1}{4}$ and $\frac{2}{8}$ are equivalent fractions. You can write $\frac{1}{4} = \frac{2}{8}$.

Fraction circles show equivalent fractions when they represent equal areas of the same-size whole.

If the red circle is the whole, then the orange fraction circle piece represents $\frac{1}{3}$.

Two light blue pieces cover the same amount of the whole as the orange piece. So, two-sixths is the same as one-third.

Two light blue pieces cover the same amount of the whole as one orange piece even when the pieces are in different places. So, two-sixths is the same as one-third.

$\frac{2}{6}$ is equivalent to $\frac{1}{3}$. Write $\frac{2}{6} = \frac{1}{3}$.

Table of Equivalent Fractions

The table below lists **equivalent fractions.** All of the fractions in a row name the same part of a whole. For every row, the red circle ● is the whole.

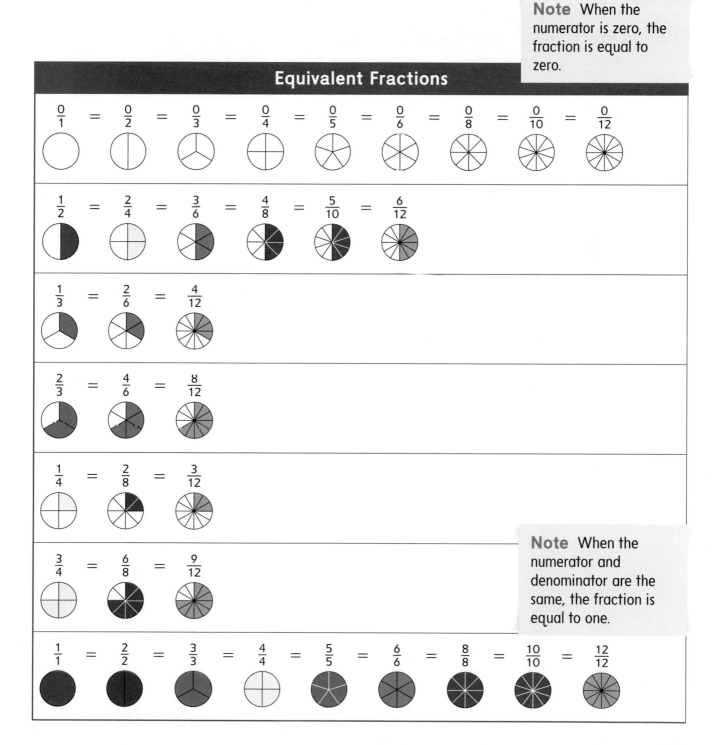

Fraction Names for Whole Numbers

You can use fraction circles, fraction strips, number lines, or drawings to represent and name whole numbers as fractions.

Examples

How many *halves* make one whole?

One-half is one out of 2 equal parts of a whole.

Count how many halves make the whole: 1-half, 2-halves.

2-halves equal one whole.

$\frac{2}{2} = 1$

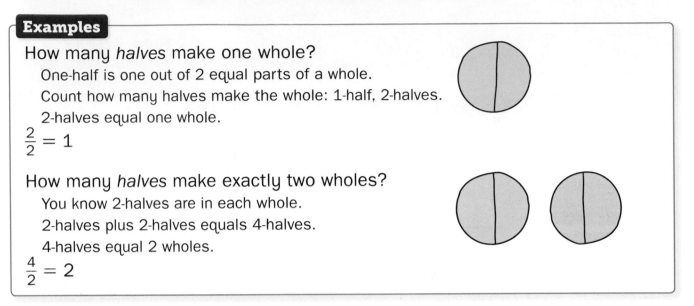

How many *halves* make exactly two wholes?

You know 2-halves are in each whole.

2-halves plus 2-halves equals 4-halves.

4-halves equal 2 wholes.

$\frac{4}{2} = 2$

Fraction number lines can help you represent whole numbers as fractions.

Example

What is a fraction that names the point at 3?

Count the wholes from 0 to 3:

0 wholes, 1 whole, 2 wholes, 3 wholes.

The fraction, 3 wholes, names the point at 3. This can be written $\frac{3}{1}$.

$\frac{3}{1}$ is equivalent to 3.

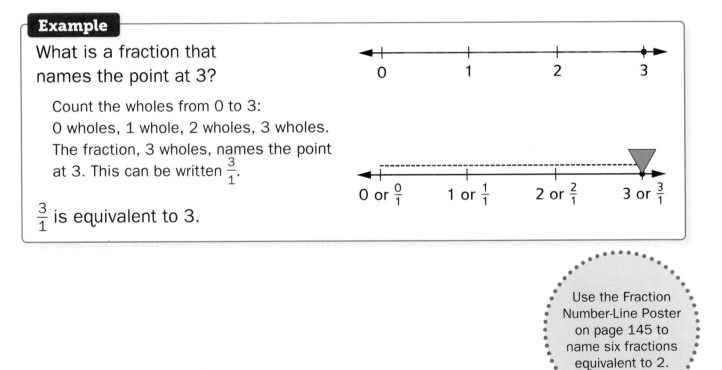

Use the Fraction Number-Line Poster on page 145 to name six fractions equivalent to 2.

Comparing Fractions

You can compare fractions when they name parts of the same whole. You can use fraction tools and pictures to compare fractions.

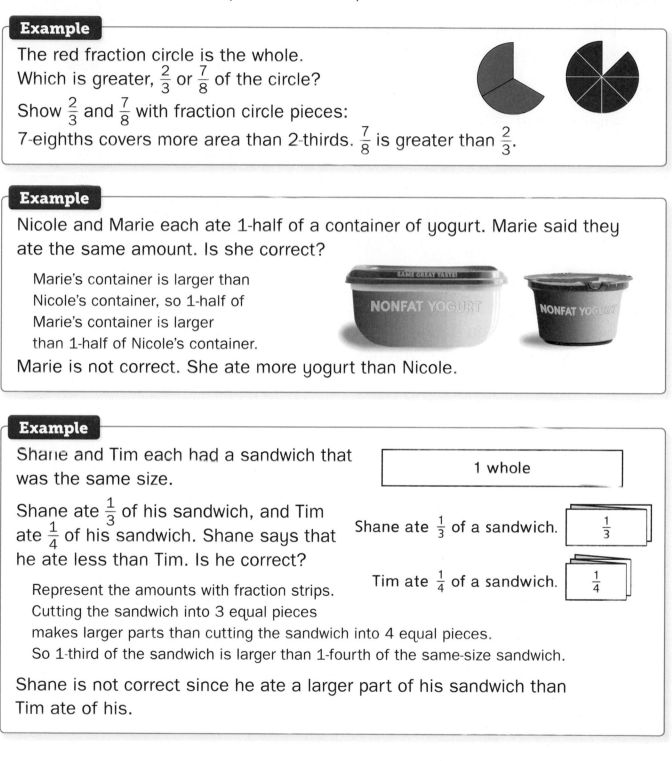

Example

The red fraction circle is the whole.
Which is greater, $\frac{2}{3}$ or $\frac{7}{8}$ of the circle?

Show $\frac{2}{3}$ and $\frac{7}{8}$ with fraction circle pieces:

7-eighths covers more area than 2-thirds. $\frac{7}{8}$ is greater than $\frac{2}{3}$.

Example

Nicole and Marie each ate 1-half of a container of yogurt. Marie said they ate the same amount. Is she correct?

Marie's container is larger than Nicole's container, so 1-half of Marie's container is larger than 1-half of Nicole's container.

Marie is not correct. She ate more yogurt than Nicole.

Example

Shane and Tim each had a sandwich that was the same size.

Shane ate $\frac{1}{3}$ of his sandwich, and Tim ate $\frac{1}{4}$ of his sandwich. Shane says that he ate less than Tim. Is he correct?

1 whole

Shane ate $\frac{1}{3}$ of a sandwich. $\frac{1}{3}$

Tim ate $\frac{1}{4}$ of a sandwich. $\frac{1}{4}$

Represent the amounts with fraction strips. Cutting the sandwich into 3 equal pieces makes larger parts than cutting the sandwich into 4 equal pieces. So 1-third of the sandwich is larger than 1-fourth of the same-size sandwich.

Shane is not correct since he ate a larger part of his sandwich than Tim ate of his.

When comparing fractions, it sometimes helps to compare each fraction to a common **benchmark,** such as 0, $\frac{1}{2}$, or 1.

Example

Compare $\frac{3}{4}$ and $\frac{3}{6}$ of the same whole. Think about a red circle as the whole.

Since $\frac{3}{4}$ is greater than 1-half, and $\frac{3}{6}$ is equal to 1-half, $\frac{3}{4}$ is greater than $\frac{3}{6}$, and $\frac{3}{6}$ is less than $\frac{3}{4}$.

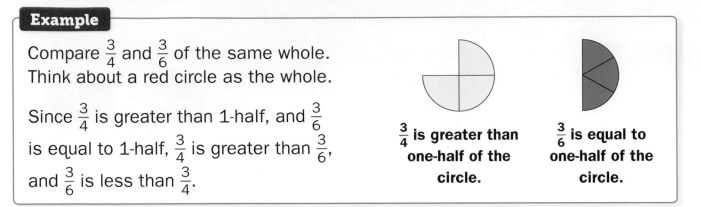

$\frac{3}{4}$ is greater than one-half of the circle.

$\frac{3}{6}$ is equal to one-half of the circle.

Sometimes you can compare fractions by thinking about which amount is closer to 1 whole.

Example

Kara and Tania each had a small pizza for lunch. The pizzas were the same size when they started eating. Kara ate $\frac{5}{6}$ of her pizza. Tania ate $\frac{3}{4}$ of her pizza. Who ate more pizza? Who has more pizza left?

You can model the pizza that each girl ate using fraction circle pieces. A red fraction circle represents the whole pizza.

The fraction circle pieces show the $\frac{5}{6}$ pizza that Kara ate. The empty part of the circle shows the piece of pizza that Kara did not eat.

The fraction circle pieces show the $\frac{3}{4}$ pizza that Tania ate. The empty part of the circle shows the piece of pizza that Tania did not eat.

The piece of pizza that Tania *did not eat* is larger than the piece of pizza that Kara *did not eat*. That means the amount that Kara ate is *closer to a whole pizza.*

Kara ate more pizza. Tania has more pizza left.

Using Symbols to Show Comparisons

When comparing two fractions, two results are possible:

- The fractions are **equal.** An *equal sign* (=) shows that the fractions represent the same amount. When fractions are equal, they are **equivalent.**

- The fractions are **not equal.** One of the fractions is larger than the other. A *greater-than symbol* (>) or a *less-than symbol* (<) shows that the fractions are not equal.

Note To remember the meaning of the > and < symbols, think of each one as a mouth. The mouth must be open wide to swallow the larger number.

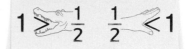

$$1 > \frac{1}{2} \quad \frac{1}{2} < 1$$

You can compare fractions using number lines.

Example

Compare $\frac{5}{8}$ and $\frac{1}{4}$.

Use number lines to think about how far each fraction is from 0.

$\frac{1}{4}$ is closer to 0, so $\frac{1}{4}$ is less than $\frac{5}{8}$.
Write $\frac{1}{4} < \frac{5}{8}$.

$\frac{5}{8}$ is farther from 0, so $\frac{5}{8}$ is greater than $\frac{1}{4}$. Write $\frac{5}{8} > \frac{1}{4}$.

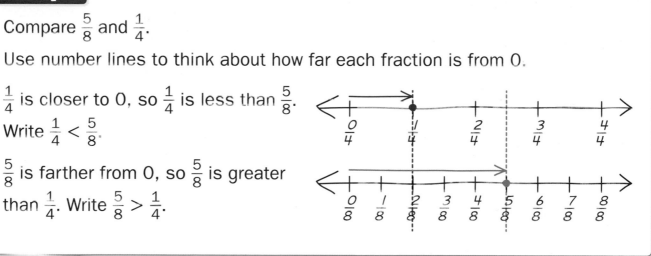

Example

Compare $\frac{4}{6}$ and $\frac{4}{3}$.

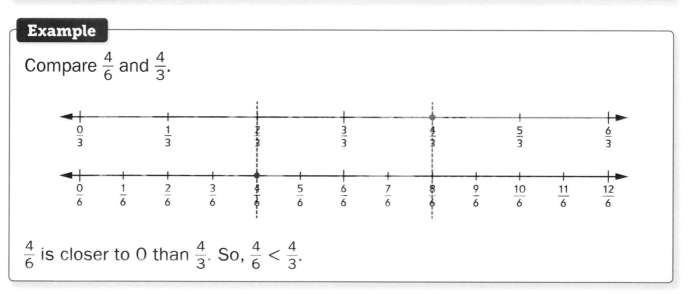

$\frac{4}{6}$ is closer to 0 than $\frac{4}{3}$. So, $\frac{4}{6} < \frac{4}{3}$.

You can use different tools or sketch your own pictures to compare fractions.

Example

Compare $\frac{2}{6}$ and $\frac{2}{8}$.

Look at the pictures on the fraction cards. 2-sixths covers more area than 2-eighths.

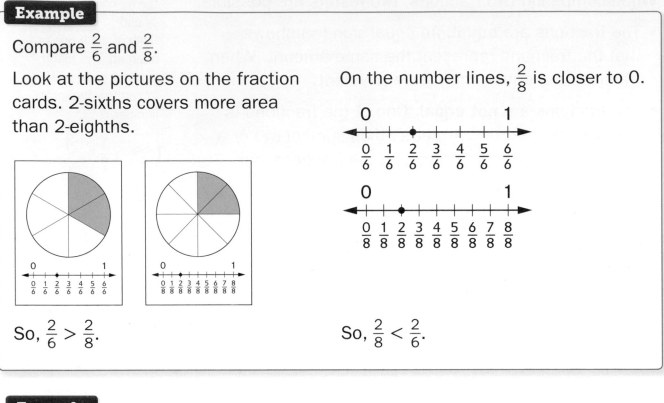

On the number lines, $\frac{2}{8}$ is closer to 0.

So, $\frac{2}{6} > \frac{2}{8}$.

So, $\frac{2}{8} < \frac{2}{6}$.

Example

Compare $\frac{3}{4}$ and $\frac{6}{8}$.

You can make a sketch of each fraction using same-size wholes.

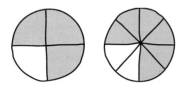

$\frac{3}{4}$ and $\frac{6}{8}$ cover the same amount of the circles.

So, $\frac{3}{4} = \frac{6}{8}$.

You can use fraction strips to compare.

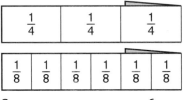

$\frac{3}{4}$ is equivalent to $\frac{6}{8}$.

Using Fractions to Describe Equal Shares

You can use fractions to figure out equal, or fair, shares of a whole. The whole can be an object, a distance, or a collection.

Examples

Rick, Sasha, and Kim share 2 bagels fairly. How much does each friend get?

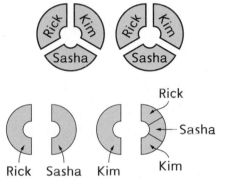

One way:

Each friend gets $\frac{1}{3}$ of each bagel. There are 2 whole bagels, so each friend gets $\frac{2}{3}$ of a bagel.
$\frac{1}{3} + \frac{1}{3} = \frac{2}{3}$

Another way:

Each friend gets 1-half of a bagel and 1-third of another half.

Example

Lola, Olive, and Sherbie get equal amounts of 5 dog treats. How many dog treats does each dog get?

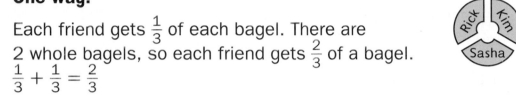 represents 1 whole dog treat.

One way: | L | O | S | L | O | S | L | O | S | L | O | S | L | O | S |

L = Lola   O = Olive   S = Sherbie

Each dog gets $\frac{1}{3}$ of each of the 5 dog treats. $\frac{1}{3} + \frac{1}{3} + \frac{1}{3} + \frac{1}{3} + \frac{1}{3} = \frac{5}{3}$

Each dog gets $\frac{5}{3}$ dog treats.

Another way:

$1 + \frac{1}{3} + \frac{1}{3} = 1\frac{2}{3}$

Each dog gets $1\frac{2}{3}$ dog treats.

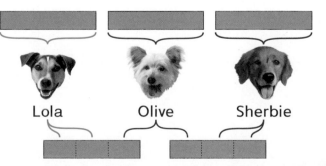

Lola   Olive   Sherbie

Example

Three small pizzas are shared equally among four people. How much pizza does each person get?

One way:

Show 3 pizzas using 3 red fraction circles.

Trade fraction circle pieces to divide each pizza into 4 equal parts.

Each person gets $\frac{1}{4}$ of each pizza. There are 3 pizzas, so each person gets $\frac{1}{4} + \frac{1}{4} + \frac{1}{4}$, or $\frac{3}{4}$, of a pizza.

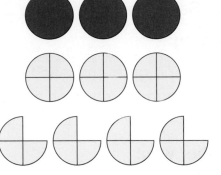

Another way:

Show 3 pizzas using 3 red fraction circles.

Trade fraction circle pieces to divide 2 of the pizzas in half. Each person gets $\frac{1}{2}$ of a pizza.

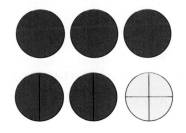

Trade fraction circle pieces to divide the last pizza into 4 equal parts. Each person gets another $\frac{1}{4}$ of a pizza.

All together, each person gets $\frac{1}{2}$ and $\frac{1}{4}$ of a pizza.

Both methods show that each person gets the same amount of a pizza.

Note If you don't have fraction circles or don't have enough for a problem, you can draw pictures or use other tools to solve fraction problems.

Solving Problems with Fractions

You can use fraction circles, fraction strips, number lines, counters, or pictures to help you solve many types of problems with fractions. Think about which tool will help you make sense of the problem.

Example

Eight friends are planting a garden. Three of the friends are planting peppers. The rest of the group is planting tomatoes. What fraction of the group is planting peppers? What fraction of the group is planting tomatoes?

You can use the eighths fraction strip. The whole strip represents the whole group of friends. Each part represents one friend, or $\frac{1}{8}$ of the group.

Peppers			Tomatoes				
$\frac{1}{8}$	$\frac{1}{8}$	$\frac{1}{8}$	$\frac{1}{8}$	$\frac{1}{8}$	$\frac{1}{8}$	$\frac{1}{8}$	$\frac{1}{8}$

Since 3 friends are planting peppers, $\frac{3}{8}$ (three-eighths) of the group are planting peppers.

There are 5 more friends in the whole group, so $\frac{5}{8}$ (five-eighths) of the group are planting tomatoes.

Example

Zharia ran $\frac{2}{3}$ mile in gym class. Derrick ran $\frac{2}{4}$ mile. Who ran farther?

You can show $\frac{2}{3}$ mile on a number line.

Divide the length between 0 and 1 into thirds.

Count: 0-thirds, 1-third, 2-thirds.
Mark $\frac{2}{3}$.

Show $\frac{2}{4}$ mile on a number line.

Divide the whole into fourths.

Count: 0-fourths, 1-fourth, 2-fourths.
Mark $\frac{2}{4}$.

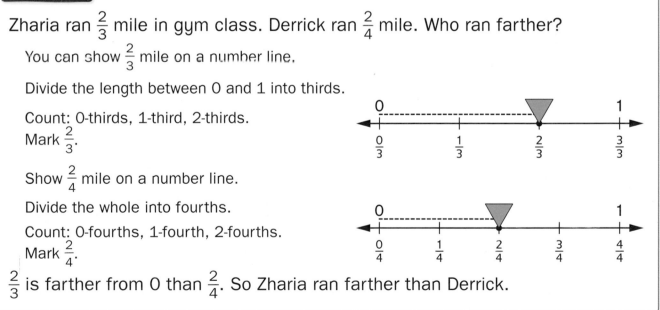

$\frac{2}{3}$ is farther from 0 than $\frac{2}{4}$. So Zharia ran farther than Derrick.

Example

Garrett has $4\frac{1}{2}$ feet of yarn for an art project. If he cuts the yarn into $\frac{1}{2}$-foot pieces, how many pieces will he have?

Draw a line segment that is 5 units long. Partition it into $\frac{1}{2}$-units. Mark $4\frac{1}{2}$.

nine $\frac{1}{2}$-foot pieces

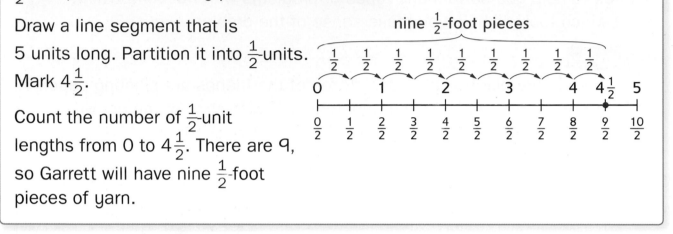

Count the number of $\frac{1}{2}$-unit lengths from 0 to $4\frac{1}{2}$. There are 9, so Garrett will have nine $\frac{1}{2}$-foot pieces of yarn.

Check Your Understanding

Solve the problems below. You can use fraction circles, fraction strips, number lines, counters, or pictures to help you.

1. Chris has a stack of 7 books. Each book is $\frac{1}{2}$ inch thick. How tall is the stack of 7 books?

2. Tucker has $\frac{3}{4}$ cup of flour. How much more flour does he need to have 1 whole cup?

3. Four loaves of bread are shared equally among 6 people. How much bread does each person get? Use a drawing to show how you solved the problem.

4. There are 3 bags of food for 2 dogs. What fraction of the bags of food does each dog get if the food is given equally? Use a drawing to show how you solved the problem.

Check your answers in the Answer Key.

Measurement and Data

Measurement Before the Invention of Standard Units

People measured length and weight long before they had rulers and scales. In the past, people used parts of their bodies to measure lengths. Here are some units of length that were based on the human body.

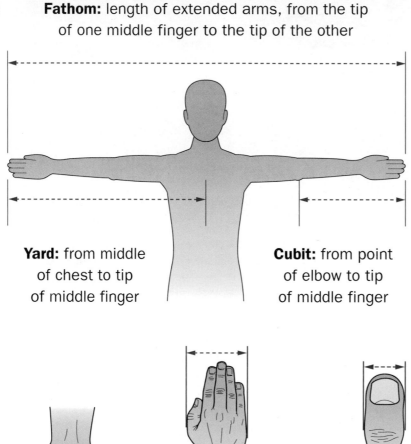

Fathom: length of extended arms, from the tip of one middle finger to the tip of the other

Yard: from middle of chest to tip of middle finger

Cubit: from point of elbow to tip of middle finger

Foot: length of foot

Hand: width of hand including thumb

Thumb: thickest width of thumb

Look carefully. You will see that a fathom is about the same length as a person's height.

The problem with using body measures is that they are different for different people.

Using **standard units** solves this problem. Standard units never change. They are the same for everyone. If two people measure the same object using standard units, their measurements will be the same or almost the same.

Check Your Understanding

Work with a partner. Each partner measures the objects below using the units in the problems. See the drawings on page 164. Are your measurements the same? Why or why not?

1. Measure the length of a desk in *hands*.

2. Measure the length of a room in *fathoms*.

Check your answers in the Answer Key.

The Metric System

About 200 years ago, the **metric system** of measurement was developed. The metric system uses standard units for measuring length, mass, and liquid volume.

- The standard unit for length is the **meter.** The word *meter* is abbreviated **m.** A meter is about the length of a big step or the width of a front door.

- The standard unit for mass is the **gram.** The word *gram* is abbreviated **g.** A dime has a mass of about 2 grams. A large paper clip has a mass of about 1 gram.

- The standard unit for liquid volume is the **liter.** The word *liter* is abbreviated **L.** The juice in four juice boxes measures about a liter.

The metric system is used all over the world for everyday measurements. Scientists almost always measure using the metric system. Metric units are often used in sports such as track and field, ice-skating, and swimming. Many food labels include metric measurements.

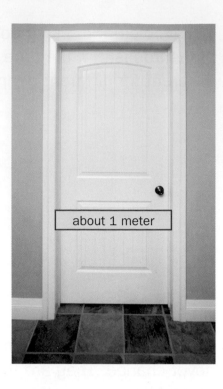

about 1 meter

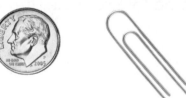

This boy is running a 100-meter race.

The mass of the cinnamon in this can is 113 grams.

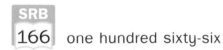

The metric system is easy to use because it is a decimal system. It is based on the numbers 10, 100, and 1,000.

You can see what this means by looking at a **meterstick.** A meterstick is a ruler that is 1 meter long.

Example

Part of a meterstick is shown below. It has been divided into smaller units.

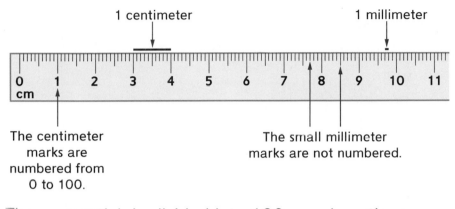

1 centimeter

1 millimeter

The centimeter marks are numbered from 0 to 100.

The small millimeter marks are not numbered.

The meterstick is divided into 100 equal sections. The length of each section is called a **centimeter.** There are 100 centimeters in 1 meter.

Each centimeter is divided into 10 equal sections. The length of each small section is called a **millimeter.** There are 10 millimeters in 1 centimeter. There are 1,000 millimeters in 1 meter.

Did You Know?

In 2003, Chinese scientists found the body of a four-winged dinosaur that was about 75 centimeters long from head to tail.

In 2011, a farm worker in Argentina accidentally discovered the bones of a Titanosaur. Scientists estimate that the length of the Titanosaur from head to tail was 40 meters.

Measuring Length in Centimeters

Length is the measure of a distance between two points. Length is usually measured with a ruler. The edges of your Pattern-Block Template are rulers. Tape measures, yardsticks, and metersticks are rulers that are used for measuring longer distances.

Rulers are often marked with **inches** on one edge and **centimeters** on the other edge. The side showing centimeters is called the centimeter scale. The side showing inches is called the inch scale.

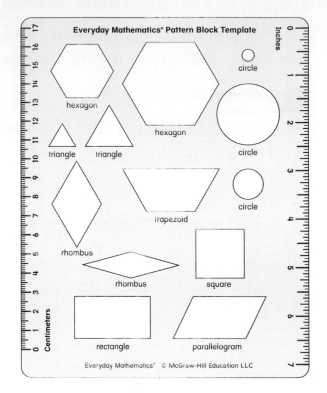

Everyday Mathematics® Pattern Block Template

hexagon
hexagon
triangle   triangle
rhombus
trapezoid
rhombus   square
rectangle   parallelogram
circle

Everyday Mathematics® © McGraw-Hill Education LLC

The centimeter marks are numbered 0, 1, 2, and so on.

centimeter scale

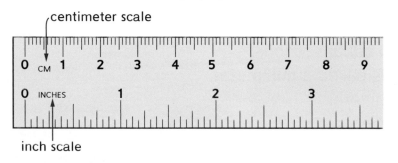

inch scale

Example

How long is the key?

You can line up one end of the object with the 0-mark on the ruler.

The other end of the key is at the 4-centimeter mark.

The distance from the 0-mark to the 4-mark on the ruler is 4 centimeters, so the key is 4 centimeters long.

Write this as 4 cm.

McGraw-Hill Education

You can measure an object by lining up one end of the object with any whole-number mark on a ruler. The length is the distance between the two ends of the object.

Example

How long is the crayon?

One end of the crayon is lined up at the 2 cm mark.

The other end of the crayon is at the 9 cm mark.

The distance between 2 cm and 9 cm is the length of the crayon. You can subtract to find the distance.

$$9 \text{ cm} - 2 \text{ cm} = 7 \text{ cm}$$

Since the distance from 2 cm to 9 cm is 7 cm, the crayon is 7 cm long.

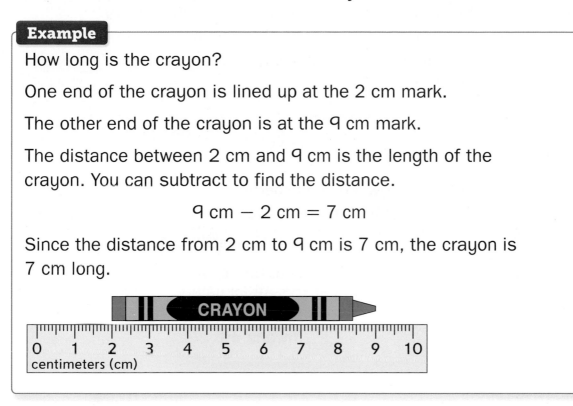

You can use a ruler to draw a line segment.

Example

Draw a line segment that is 7 centimeters long.

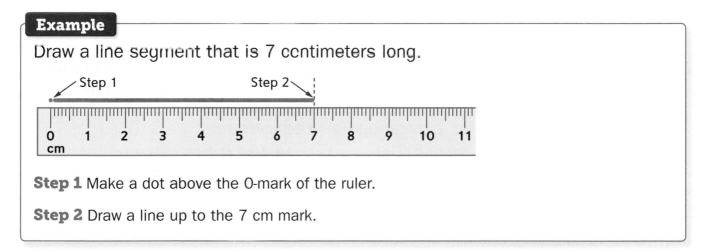

Step 1 Make a dot above the 0-mark of the ruler.

Step 2 Draw a line up to the 7 cm mark.

Personal References for Metric Units of Length

Sometimes you may not have a ruler or meterstick handy. When this happens, you can estimate lengths by using the lengths of common objects and distances you know. These are called *personal references*. Some examples are given below.

Personal References for Metric Units of Length	
About 1 millimeter	**About 1 centimeter**
• thickness of a pushpin point • thickness of a dime	• width of a fingertip • thickness of a crayon
About 1 meter	**About 1 kilometer**
• one big step (for an adult) • width of a front door	• 1,000 big steps (for an adult) • length of 10 football fields (including the end zones)

The point of the pushpin is about 1 millimeter thick.

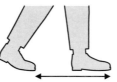

One big step for an adult is about 1 meter long.

1,000 big steps for an adult is about 1 kilometer long.

The width of your finger is about 1 centimeter.

The personal references for 1 meter can also be used for 1 yard. One meter is slightly longer than 39 inches. One yard equals 36 inches. So a meter is about 3 inches longer than a yard.

meterstick

```
0  2  4  6  8  10 12 14 16 18 20 22 24 26 28 30 32 34 36 38 40 42 44 46 48 50 52 54 56 58 60 62 64 66 68 70 72 74 76 78 80 82 84 86 88 90 92 94 96 98 100
```

```
0  1  2  3  4  5  6  7  8  9  10 11 12 13 14 15 16 17 18 19 20 21 22 23 24 25 26 27 28 29 30 31 32 33 34 35 36
```

yardstick

University of Chicago

Measuring Length in Inches

Length is the measure of a distance between two points. In the U.S. customary system, a standard unit of length is the **inch.** The word *inch* is abbreviated **in.**

On rulers, inches are usually divided into halves, quarters (or fourths), eighths, and sixteenths. The marks to show fractions of an inch are usually different sizes.

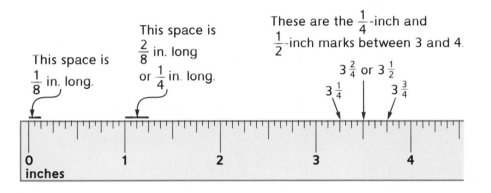

This space is $\frac{1}{8}$ in. long.

This space is $\frac{2}{8}$ in. long or $\frac{1}{4}$ in. long.

These are the $\frac{1}{4}$-inch and $\frac{1}{2}$-inch marks between 3 and 4.

$3\frac{1}{4}$ $3\frac{2}{4}$ or $3\frac{1}{2}$ $3\frac{3}{4}$

Example

What is the length of the eraser?

You can line up the end of the object with the 0-mark of the ruler. The distance from the 0-mark on the ruler to the 2-inch mark is the length of the eraser.

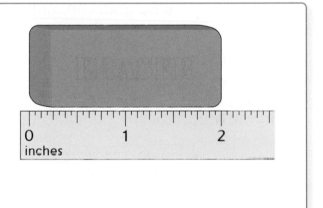

The eraser is about 2 inches long.

Write this as 2 in.

If the 0-mark is at the end of a ruler, the number 0 may not be printed on that ruler.

You can measure an object by lining up one end of the object with any mark on a ruler. The length is the distance between the two ends of the object.

Example

How long is the marker?

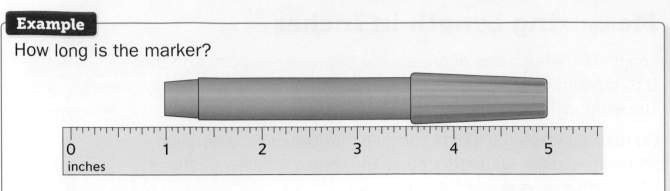

One end of the marker is lined up at the 1 in. mark. The other end of the marker is at the 5 in. mark.

The distance between 1 in. and 5 in. is the length of the marker. You can count up to find the distance.

1 in. + **?** = 5 in.

1 in. + **4 in.** = 5 in.

Since the distance from 1 in. to 5 in. is 4 in., the marker is 4 in. long.

Decide how **precise** your measurement will be. For example, you can measure to the nearest $\frac{1}{4}$ inch or to the nearest $\frac{1}{2}$ inch. You may decide that measuring to the nearest inch is close enough.

Example

Find the length of the pencil to the nearest quarter-inch and nearest inch.

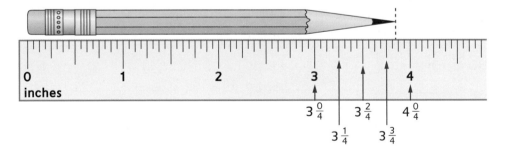

The quarter-inch marks between 3 and 4 are written below the ruler. The tip of the pencil is closest to the $3\frac{3}{4}$-inch mark.

The pencil is $3\frac{3}{4}$ inches long, to the nearest quarter-inch.

The tip of the pencil is closer to the 4-inch mark than the 3-inch mark.

The pencil is 4 inches long, to the nearest inch.

Personal References for U.S. Customary Units of Length

Sometimes you may not have a ruler or yardstick handy. When this happens, you can estimate lengths by using the lengths of common objects and distances you know. These are called *personal references*. Some examples are given below.

Personal References for U.S. Customary Units of Length	
About 1 inch	**About 1 foot**
• width of a quarter • width of a man's thumb	• distance from elbow to wrist (for an adult) • length of a piece of paper
About 1 yard	**About 1 mile**
• one big step (for an adult) • width of a front door	• length of 15 football fields (including the end zones) • 2,000 big steps (for an adult)

The width of a quarter is about 1 inch.

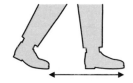

One big step for an adult is about 1 yard.

2,000 big steps for an adult is about 1 mile.

The distance from an adult's elbow to wrist is about 1 foot.

The personal references for 1 meter can also be used for 1 yard. One meter is slightly longer than 39 inches. One yard equals 36 inches. So a meter is about 3 inches longer than a yard.

meterstick

```
0  2  4  6  8  10 12 14 16 18 20 22 24 26 28 30 32 34 36 38 40 42 44 46 48 50 52 54 56 58 60 62 64 66 68 70 72 74 76 78 80 82 84 86 88 90 92 94 96 98 100
```

```
0  1  2  3  4  5  6  7  8  9  10 11 12 13 14 15 16 17 18 19 20 21 22 23 24 25 26 27 28 29 30 31 32 33 34 35 36
```

yardstick

Perimeter

Sometimes you want to know the **perimeter,** or distance around a shape. To measure a perimeter, use units of length such as inches, meters, or miles.

Example

Sophia rode her bicycle once around the edge of a lake.

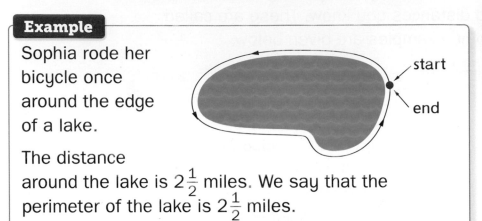

The distance around the lake is $2\frac{1}{2}$ miles. We say that the perimeter of the lake is $2\frac{1}{2}$ miles.

The inside edge of the sidewalk creates a boundary around each section of grass. The length of this boundary is the perimeter of a section.

To find the perimeter of any polygon, you can add the length of each of its sides. Always remember to name the unit of length used to measure the polygon.

Example

Alan and his friends want to build a fence around a rectangular field. How much fencing is needed to enclose the field?

The amount of fencing needed is the perimeter of the field. To find the amount needed to enclose the field, add the lengths of all four sides.

Alan and his friends need 320 yards of fencing.

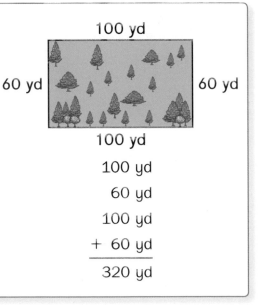

100 yd

60 yd   60 yd

100 yd

100 yd
100 yd
60 yd
100 yd
+ 60 yd
─────
320 yd

Peter Kirillov/iStock/360/Getty Images

Example

Find the perimeter of this square.

All four sides have the same length.

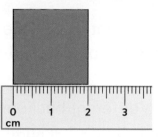

The picture shows that one side is 2 centimeters long.

Add the lengths of the four sides.

2 cm + 2 cm + 2 cm + 2 cm = 8 cm

The perimeter of the square is 8 centimeters.

Did You Know?

The perimeter of the United States, not counting Alaska and Hawaii, is about 11,300 miles. That means that if you walked around the boundary of the United States, you would walk more than 11,000 miles.

Check Your Understanding

Find the perimeter of the polygons below.

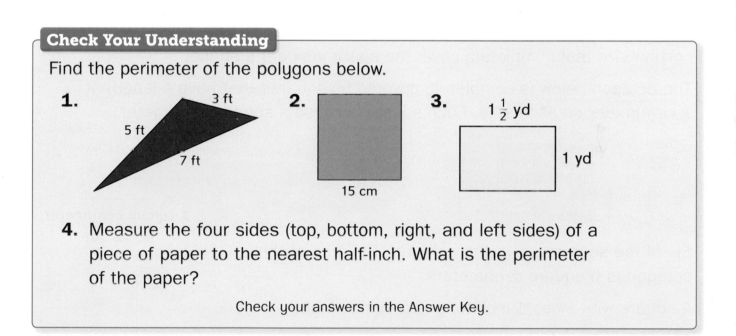

1. 3 ft, 5 ft, 7 ft

2. 15 cm

3. $1\frac{1}{2}$ yd, 1 yd

4. Measure the four sides (top, bottom, right, and left sides) of a piece of paper to the nearest half-inch. What is the perimeter of the paper?

Check your answers in the Answer Key.

Area

Sometimes we want to know the amount of surface inside a shape. The amount of surface inside the boundary of a closed shape is called the **area** of a shape.

The drawings below represent two different backyards. Carla's backyard has a larger area than Greg's backyard. Greg's backyard could be placed inside of Carla's backyard with space left over.

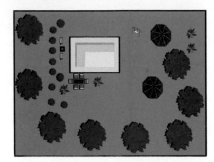

Carla's backyard

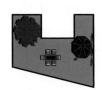

Greg's backyard

Sometimes we need more precise ways of finding the areas of shapes. One way to find the area of a shape is to count the number of squares of a certain size that completely cover the space inside the shape.

The octagon below is completely covered by squares that have a length of 1 centimeter on each side. Each square is called a **square centimeter.**

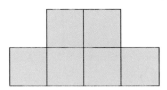

1 cm

1 cm

1 square centimeter
(actual size)

Six of the squares cover the octagon. The area of the octagon is 6 square centimeters.

A square with sides 1 inch long is a **square inch.**

A square with sides 1 foot long is a **square foot.**

The **square yard** and **square meter** are larger units of area. They are used to measure large areas such as the area of a floor.

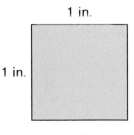

1 in.

1 in.

1 square inch
(actual size)

You can count square units to find the areas of shapes.

Example

Count the square units to find the areas of these shapes.

Each square is 1 square inch.
18 squares cover the rectangle.

The area of the rectangle is 18 square inches.

Each square is 1 square foot.
14 squares cover the shape.

The area of the shape is 14 square feet.

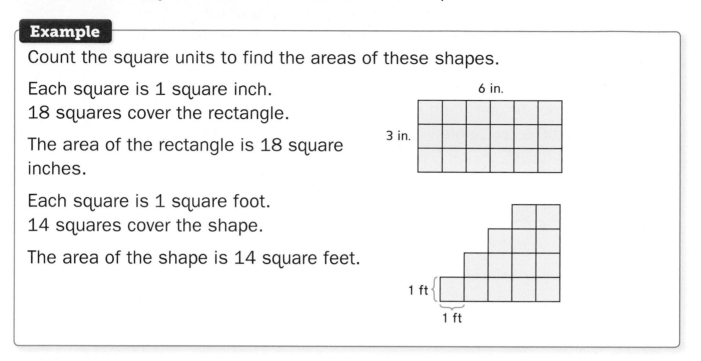

6 in.

3 in.

1 ft

1 ft

Remember: Perimeter is the **distance around** a shape.

Area is the amount of **surface inside** the shape.

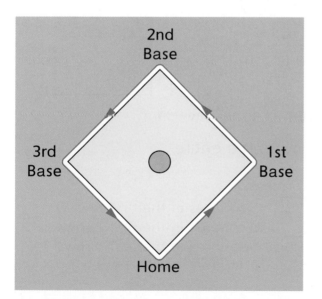

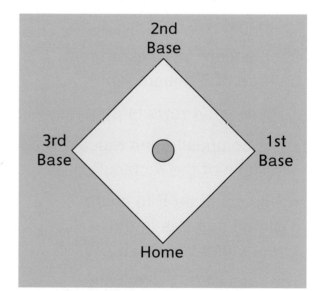

The distance between bases on youth baseball and softball fields is 60 feet. The base paths form a boundary around the infield. The perimeter, or the distance around the infield, is 60 + 60 + 60 + 60 = 240 feet.

The area of the infield, or the surface inside the base paths, is 360 square feet. That means that you could draw 360 squares with side lengths of 1 foot on the infield.

Sometimes it is more efficient to find the area of the space inside a shape by counting groups of units. Smaller units can be grouped together into larger **composite units.** In the rectangle below, five squares with 1-foot sides are grouped together to form a composite unit that has an area of 5 square feet. The composite unit is shaded in red. It is one row of squares inside the rectangle.

Each square is 1 square foot.

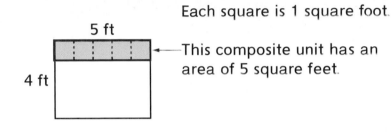

←—This composite unit has an area of 5 square feet.

You can find the area of a shape if you know the number of times you repeat the composite units as you cover the inside of a shape.

Example

Use the shaded composite unit to find the area of the rectangle.

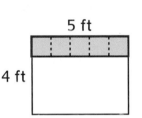

Each square is 1 square foot. The composite unit has an area of 5 square feet.

- There are 4 rows of squares, with 5 squares in each row.

- The composite unit repeats 4 times as it covers the entire surface of the rectangle.

- Skip count by 5 to count all the squares in the rectangle: 5, 10, 15, 20.

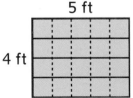

5 squares in the first row
10 squares in the first 2 rows
15 squares in the first 3 rows
20 squares in all 4 rows

The area of this rectangle is 20 square feet.

Some surfaces are too large to cover with squares. It would take a long time to count a large number of squares.

To find the area of a rectangle, you do not need to count all of the squares that cover it. The example below shows a more efficient way to find the area.

Example

Find the area of this rectangle.

Each square is 1 square foot.

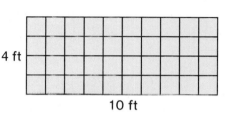

4 ft

10 ft

• There are 4 rows of squares.

• Each row has 10 squares.

• So there are 4 × 10 squares, or 40 squares in all.

The area is 40 square feet.

Summary   To find the area of a rectangle:

Step 1 Count the number of rows.

Step 2 Count the number of squares in each row.

Step 3 Multiply: (number of rows) × (number of squares in each row)

Check Your Understanding

Find the area of each rectangle.

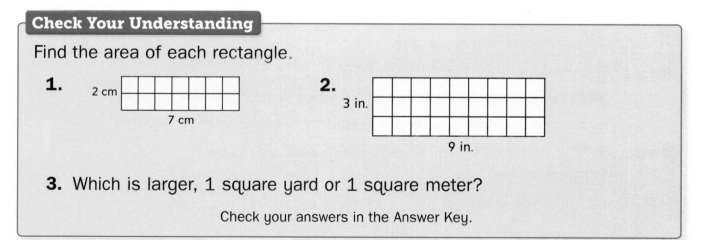

1. 2 cm

7 cm

2. 3 in.

9 in.

3. Which is larger, 1 square yard or 1 square meter?

Check your answers in the Answer Key.

The shapes below are called **rectilinear figures** because their sides are all line segments and the inside and outside corners of each shape are all right angles. Rectangles are rectilinear figures. But the figures on this page are not rectangles because they have more than four sides.

You cannot find the areas of these figures by multiplying the number of squares in each row by the number of rows because the rows do not all have the same number of squares. You can find the area of a rectilinear figure by **decomposing,** or separating, it into non-overlapping rectangles, finding the area of each rectangle, and then adding the areas of these smaller rectangles.

Note You can find other ways to decompose the shape in the example below into two or more rectangles. Then follow the same steps.

Example

Find the area of this rectilinear shape. Each square is 1 square inch.

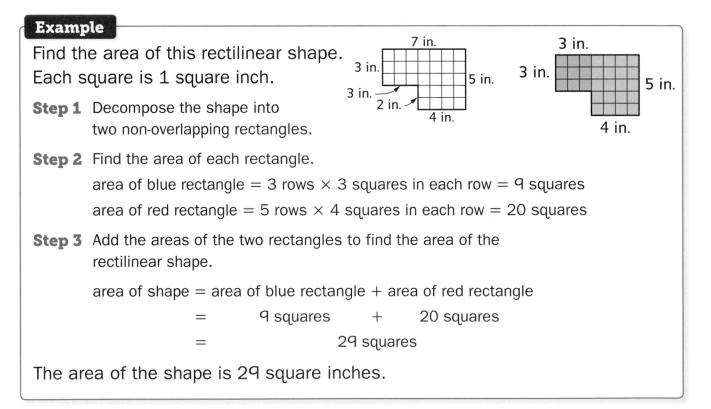

Step 1 Decompose the shape into two non-overlapping rectangles.

Step 2 Find the area of each rectangle.

area of blue rectangle = 3 rows × 3 squares in each row = 9 squares

area of red rectangle = 5 rows × 4 squares in each row = 20 squares

Step 3 Add the areas of the two rectangles to find the area of the rectilinear shape.

area of shape = area of blue rectangle + area of red rectangle

= 9 squares + 20 squares

= 29 squares

The area of the shape is 29 square inches.

You can also look at the rectilinear shape in the example on the previous page as a large rectangle with the corner cut out. Another way to find the area is to first find the area of the large rectangle, and then subtract the area of the small corner rectangle.

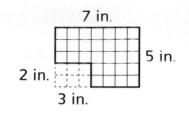

7 in.

5 in.

2 in.

3 in.

area of rectilinear shape	=	area of large rectangle	–	area of small corner rectangle
	=	(5 rows × 7 squares in each row)	–	(2 rows × 3 squares in each row)
	=	35 squares	–	6 squares
	=		29 squares	

The area of the rectilinear shape is 29 square inches.

Finding the area of a shape using two different methods is a good way to check your work. Both methods show that the area of the rectilinear shape is 29 square inches.

Check Your Understanding

1. Draw the shape in the example on the previous page on grid paper or make a sketch. Decompose it into different rectangles than those shown in the example. Find the sum of their areas. Did you find the same area as the area in the example?

2. Copy the figure on the right on grid paper or make a sketch. Find its area by decomposing it into rectangles.

3. Find the area of the rectilinear shape by decomposing it into rectangles.

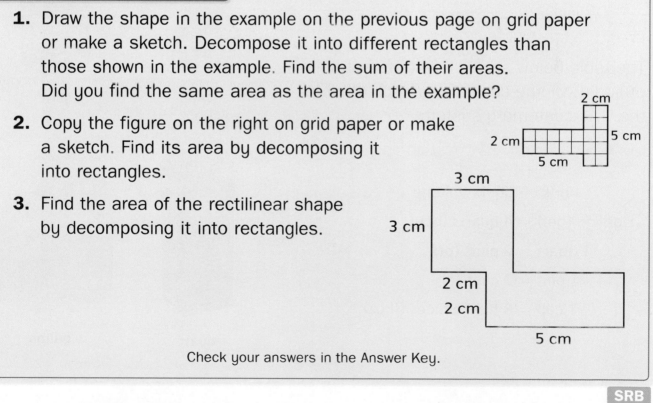

2 cm

2 cm

5 cm

5 cm

3 cm

3 cm

2 cm

2 cm

5 cm

Check your answers in the Answer Key.

Liquid Volume

Sometimes you need to know amounts of things that can be poured. All liquids can be poured. Some solids, such as sand and sugar, can be poured, too.

Liquid volume is a measure of how much liquid is in a container. Liquid volume is usually measured in units such as **liters, milliliters, gallons, quarts, pints, cups,** and **fluid ounces.**

Liters and milliliters are units in the **metric system.** Gallons, quarts, pints, cups, and fluid ounces are units in the **U.S. customary system.** Most labels for liquid containers show volume in both metric and U.S. customary units.

The table below shows how different units of liquid volume compare to each other in the metric system.

1,000 mL beaker, or a 1 L beaker

A beaker is a tool that measures liquid volume.

Metric Units
Units of Liquid Volume
1 liter (L) = 1,000 milliliters (mL)
1 milliliter = $\frac{1}{1,000}$ liter

The table below shows how different units of liquid volume compare to each other in the U.S. customary system.

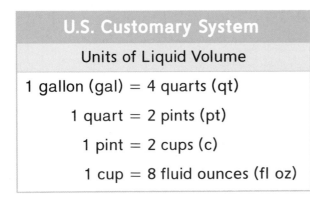

cup   pint

U.S. Customary System
Units of Liquid Volume
1 gallon (gal) = 4 quarts (qt)
1 quart = 2 pints (pt)
1 pint = 2 cups (c)
1 cup = 8 fluid ounces (fl oz)

quart   gallon

(t)DNY59/iStock/360/Getty Images, (c, cr, bcr) McGraw-Hill Education/Mark Steinmetz, (br)Mark Steinmetz

Measurement and Data

Mass

Mass is a measure of the amount of matter (solid, liquid, or gas) in an object. The standard unit for mass in the metric system is the **gram.** Another metric unit of measure for mass is the **kilogram.**

A pan balance is an example of a tool used to measure mass.

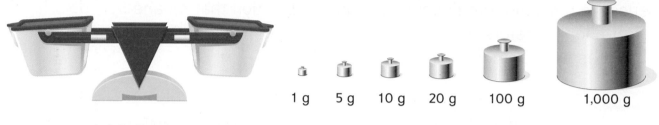

pan balance

1 g   5 g   10 g   20 g   100 g   1,000 g

masses

The table to the right shows how different units of mass compare to each other in the metric system.

You can estimate masses by using the masses of common objects that you know. These are called personal references. Some examples of personal references for mass are given below.

Metric Units
Units of Mass
1 gram (g) = 1,000 milligrams (mg)
1 kilogram (kg) = 1,000 grams

Personal References for Metric Units of Mass	
About 1 gram	About 1 kilogram
• dollar bill	• wooden baseball bat
• large paper clip	• 200 U.S. nickels

A dollar bill has a mass of about 1 gram.

A wooden baseball bat has a mass of about 1 kilogram.

One kilogram is equal to 1,000 grams. So 1,000 dollar bills would have about the same mass as a wooden bat.

Time

We use time in two ways:

1. to tell when something happens, and

2. to tell how long something takes or lasts.

A.M. is an abbreviation that means "before noon." It refers to the period from midnight to noon. P.M. is an abbreviation that means "after noon." It refers to the period from noon to midnight. Noon is written as 12:00 P.M. Midnight is written as 12:00 A.M.

Example

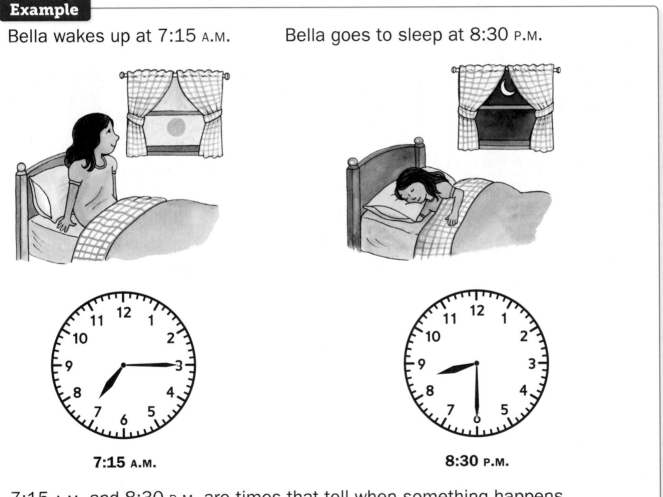

Bella wakes up at 7:15 A.M.

Bella goes to sleep at 8:30 P.M.

7:15 A.M.

8:30 P.M.

7:15 A.M. and 8:30 P.M. are times that tell when something happens.

7:15 A.M. is in the morning (before noon).

8:30 P.M. is in the evening (after noon).

The table below shows how units of time compare.

Units of Time
1 minute = 60 seconds
1 hour = 60 minutes
1 day = 24 hours
1 week = 7 days
1 month = 28, 29, 30, or 31 days
1 year = 12 months
1 year = 52 weeks plus 1 day, or 52 weeks plus 2 days in leap years
1 year = 365 days, or 366 days in leap years
1 decade = 10 years
1 century = 100 years
1 millennium = 1,000 years

Did You Know?

A Galapagos tortoise has a life span of more than 200 years.

The life span of a worker bee is between 1 and 4 months.

You can estimate amounts of time by thinking about the lengths of different activities that you do. Some examples are listed below.

- about 1 minute: singing a song like the "Alphabet Song"
- about 15 minutes: recess or playing a few rounds of a math game
- about 30 minutes: watching a television show
- about 45 minutes: art or gym class
- about 1 hour: mathematics class
- about 2 hours: watching a movie

Try timing classroom routines and activities throughout the day and add to this list.

Telling Time to the Nearest Minute

Sometimes you need a more precise, or exact, time to tell when something happens or how long something lasts. You may need to catch a bus at a certain time or know how long to cook something. Instead of telling time to the nearest hour or half hour, you may need to tell time to the nearest minute.

You can use analog clocks to tell time to the nearest minute.

Example

What time does the clock show?

The hour hand (short hand) is pointing a little past the 8-hour mark but before the 9-hour mark. The hour is 8:00.

The minute hand (long hand) is pointing between 4 and 5. Count the number of minute marks from the top of the clock to find the minutes past the hour. Count by groups of 5 until you get to the 4-mark: 5, 10, 15, 20 minutes.

The minute hand is pointing 3 minute marks past the 4. Count 3 more minutes: 21, 22, 23 minutes.

The time to the nearest minute is 8:23.

You can use familiar times to figure out time to the nearest minute.

Example

What time does the clock show?

The hour hand (short hand) is pointing past the 4-hour mark. The minute hand (long hand) is pointing a little more than halfway around the clock face.

The familiar time is 4:30. The minute hand is pointing 2 minute marks past the 30-minute mark. Count 2 more minutes.

The time to the nearest minute is 4:32.

Finding the Length of Day

Sometimes you may want to know how much time has passed. This is called **elapsed time.** The elapsed time for the **length of day** is the total number of hours and minutes between sunrise and sunset. That gives the number of hours and minutes of daylight.

You can use a toolkit clock to find the length of day. Count the hours and minutes and record your work in a table like the one in the example below.

Example

What is the length of day?

Time of Sunrise	Time of Sunset
6:22 A.M.	7:29 P.M.

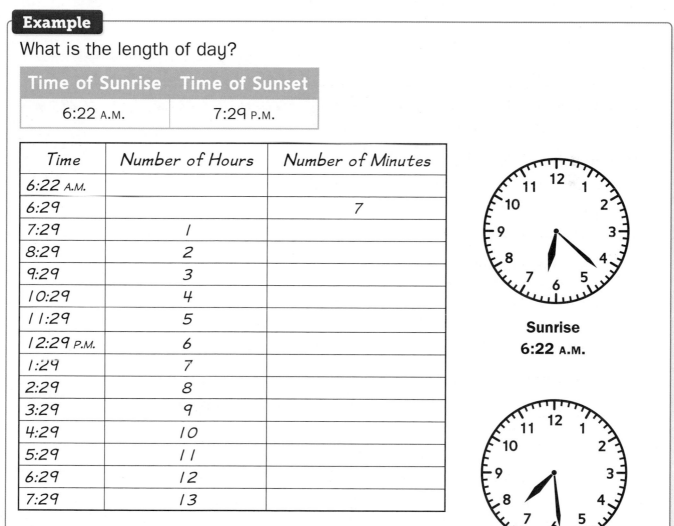

Time	Number of Hours	Number of Minutes
6:22 A.M.		
6:29		7
7:29	1	
8:29	2	
9:29	3	
10:29	4	
11:29	5	
12:29 P.M.	6	
1:29	7	
2:29	8	
3:29	9	
4:29	10	
5:29	11	
6:29	12	
7:29	13	

Sunrise
6:22 A.M.

Sunset
7:29 P.M.

There are 7 minutes from 6:22 A.M. to 6:29 A.M.
There are 13 hours from 6:29 A.M. to 7:29 P.M.

The length of day from 6:22 A.M. to 7:29 P.M. is 13 hours and 7 minutes.

You can also use an open number line to find the elapsed time for the length of day.

Example

What is the length of day?

Time of Sunrise	Time of Sunset
6:22 A.M.	7:29 P.M.

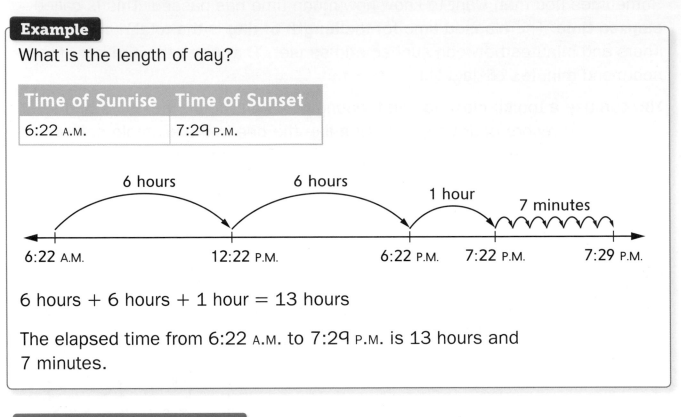

6 hours + 6 hours + 1 hour = 13 hours

The elapsed time from 6:22 A.M. to 7:29 P.M. is 13 hours and 7 minutes.

Check Your Understanding

1. How many hours pass from 7:00 A.M. to 7:00 P.M.?
   How many hours pass from 6:22 A.M. to 6:22 P.M.?
   How many hours have elapsed from 3:26 A.M. to 3:26 P.M.?
   Describe the pattern.

2. What is the longest possible time elapsed for an event if the time is within the same day?

   Check your answers in the Answer Key.

Using 24-Hour Notation

In the United States, Canada, and a few other countries, 12-hour notation is used to tell time. The 24 hours of the day are divided into two 12-hour periods. The first 12 hours of the day are A.M. times (12:00 A.M. to 11:59 A.M.). The second 12 hours of the day are P.M. times (12:00 P.M. to 11:59 P.M.)

However, in many countries of the world, 24-hour notation is used to tell time. The 24 hours of the day are numbered 0–23 to tell the number of hours passed since midnight. Another name for 24-hour notation is *military time*.

The table shows hours using 12-hour notation and 24-hour notation. In both notations, a colon is usually used to separate the hours from the minutes. Two digits before the colon tell the number of hours that have passed since midnight. A zero is usually used in front of one-digit hours. Two digits after the colon tell the number of minutes that have passed since the hour.

Time Equivalence	
12-hour notation	24-hour notation
12:00 A.M. (midnight)	00:00
1:00 A.M.	01:00
2:00 A.M.	02:00
3:00 A.M.	03:00
4:00 A.M.	04:00
5:00 A.M.	05:00
6:00 A.M.	06:00
7:00 A.M.	07:00
8:00 A.M.	08:00
9:00 A.M.	09:00
10:00 A.M.	10:00
11:00 A.M.	11:00
12:00 P.M. (noon)	12:00
1:00 P.M.	13:00
2:00 P.M.	14:00
3:00 P.M.	15:00
4:00 P.M.	16:00
5:00 P.M.	17:00
6:00 P.M.	18:00
7:00 P.M.	19:00
8:00 P.M.	20:00
9:00 P.M.	21:00
10:00 P.M.	22:00
11:00 P.M.	23:00

Example

24-hour notation
19:30

The first two digits are 1 and 9, so 19 hours have passed since midnight.

The last two digits are 3 and 0, so 30 minutes have passed since the nineteenth hour.

19:30 is read "nineteen thirty."

19:30 in 24-hour notation is the same as 7:30 P.M. in 12-hour notation.

Tally Charts

There are different ways you can collect information about something:

- count

- measure

- ask questions

- look at something and describe what you see

The information you collect is called **data.** Sometimes you can use a **tally chart** to record and organize data.

Example

Mr. Davis asked each student to name his or her favorite drink. He recorded the students' choices in the tally chart below.

Favorite Drinks

Drink	Tallies
Milk	~~IIII~~
Chocolate milk	III
Soft drink	~~IIII~~ ~~IIII~~ I
Apple juice	III
Tomato juice	I
Water	II

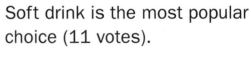

Milk (5 votes) is more popular than chocolate milk (3 votes).

Soft drink is the most popular choice (11 votes).

Tomato juice is the least popular choice (1 vote).

Check Your Understanding

1. Use the Favorite Drinks tally chart on this page to answer the questions.

a. Which drinks are less popular than apple juice?

b. How many more children chose milk than chocolate milk?

c. How many fewer children voted for apple juice than soft drink?

d. What is the total number of children who voted?

Check your answers in the Answer Key.

Bar Graphs

A **bar graph** is a drawing that uses bars to represent data. Bar graphs can help you answer questions about the data. The example below is a **scaled bar graph.** The scale shows intervals of 2.

Example

The bar graph below shows how many children in a class chose certain foods as their favorites.

The title shows the subject of the graph.

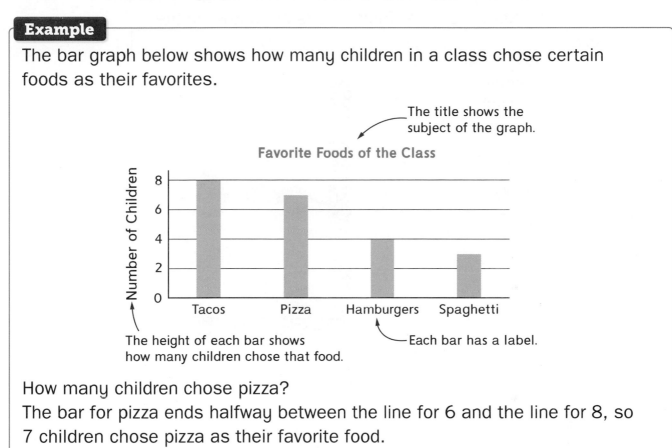

Favorite Foods of the Class

The height of each bar shows how many children chose that food.

Each bar has a label.

How many children chose pizza?
The bar for pizza ends halfway between the line for 6 and the line for 8, so 7 children chose pizza as their favorite food.

How many more children chose tacos than spaghetti?
Eight children chose tacos as their favorite food, but only 3 children chose spaghetti. Five more children chose tacos than spaghetti.

Often, you choose the scale for a bar graph based on the data and the amount of available space for the graph. If the numbers in your data set are spread out, you will want to use larger intervals to create your bar graph.

After collecting data, you can organize it in a tally chart to help you make a bar graph.

Example

The children in Mr. Majumdar's class counted how many pull-ups each of them could do. Their results are shown in the tally chart.

Number of Pull-Ups	Number of Children
0	~~HHH~~ /
1	~~HHH~~
2	////
3	//
4	
5	///
6	/

The bar graph below shows the same information as the tally chart, but in a different way.

Pull-Ups by Third Graders

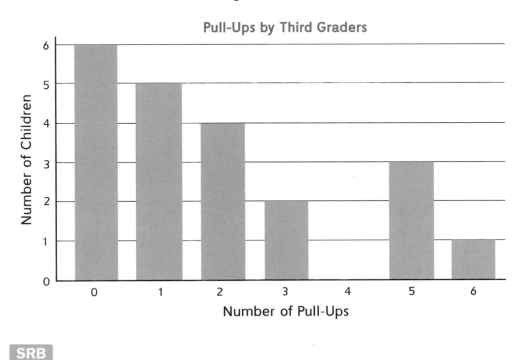

Fuse/Getty Images

Picture Graphs

A **picture graph** is a graph made with symbols. The KEY tells you how many things each symbol represents.

Example

The picture graph below shows how many children chose certain foods as their favorite.

Favorite Foods of the Class

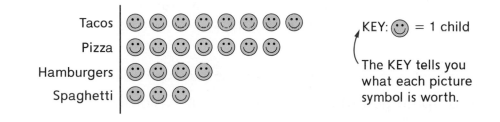

The line for tacos shows 8 face symbols.
Each face symbol stands for 1 child.
So 8 children chose tacos as their favorite food.

Scaled Picture Graphs

In a **scaled picture graph,** each symbol stands for more than one thing. In the example on the next page, each symbol stands for 10 children.

Decide on a scale for your graph by looking at your data. How much room do you have to show your data? How small is the smallest (minimum) number in your data set? How large is the largest (maximum) number in your data set? How many pictures do you want to use to represent that number?

Example

This scaled picture graph shows how many children in Lincoln School are in each grade.

Number of Children in Each Grade

3rd Grade	☺ ☺ ☺ ☺ ☺ ☺ ☺ ☺ ☺
4th Grade	☺ ☺ ☺ ☺ ☺ ☺ ☺
5th Grade	☺ ☺ ☺ ☺ ☺ ☺ ☺ ☺

KEY: ☺ = 10 children

The line for 3rd grade shows 9 face symbols. Each face symbol stands for 10 children. So there are 9 × 10, or 90, children in the 3rd grade at Lincoln School.

In some scaled picture graphs, you may see only part of a picture symbol. Use the KEY to decide how much this part of the symbol is worth.

Example

This picture graph shows how many children in each grade at Lincoln School ride a bicycle to school.

Number of Children Who Ride a Bicycle to School

KEY: ✷ = 2 children

✷ stands for 2 children, so ◖ stands for 1 child.

Check Your Understanding

Use the Number of Children Who Ride a Bicycle to School picture graph to answer the questions.

1. How many children in 3rd grade ride a bicycle to school?

2. How many more children in 5th grade ride a bicycle to school than children in 4th grade?

Check your answers in the Answer Key.

Line Plots

A **line plot** uses check marks, Xs, or other marks to show counts. There is one mark for each count.

Example

The children in Ms. Jackson's class got the scores below on a five-word spelling test. Each score is the number of words a child spelled correctly.

5 3 5 0 4 4 5 4 4 4 2 3 4 5 3 5 4 3 4 4

They drew a line plot to show the data.

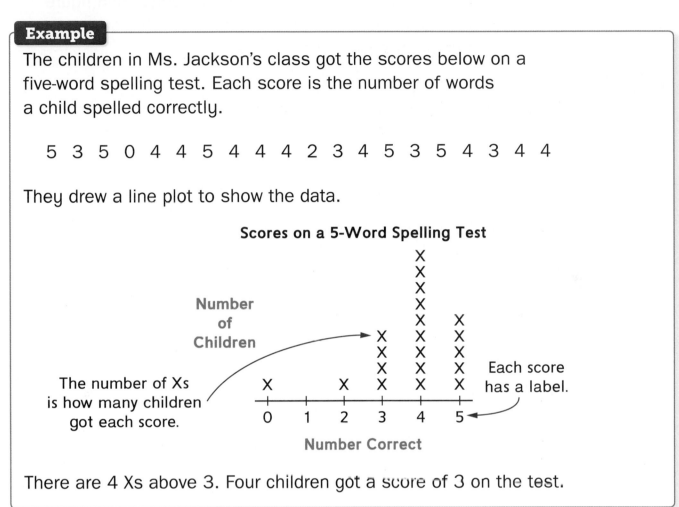

Scores on a 5-Word Spelling Test

Number of Children

The number of Xs is how many children got each score.

Each score has a label.

Number Correct

There are 4 Xs above 3. Four children got a score of 3 on the test.

You can get a lot of information from the line plot.

• The most common score on the test was 4.

• All but 2 children spelled 3 or more words correctly.

Scaled Line Plots

You can figure out the scale of a line plot by looking at a set of data. Identify the smallest (minimum) and largest (maximum) numbers in your data set. These can be the first and last numbers on your scale. Then figure out a way to represent the rest of your data. The numbers on your scale should have the same interval, or space, between them.

Example

Ashley and her classmates measured the length of their pencils to the nearest $\frac{1}{2}$ inch. When they put their measurements in order from smallest to largest, these were their results:

$4 \quad 5 \quad 5 \quad 5 \quad 5\frac{1}{2} \quad 5\frac{1}{2} \quad 5\frac{1}{2} \quad 5\frac{1}{2} \quad 6 \quad 6 \quad 6 \quad 6 \quad 6 \quad 6 \quad 6\frac{1}{2} \quad 6\frac{1}{2} \quad 6\frac{1}{2} \quad 6\frac{1}{2} \quad 6\frac{1}{2} \quad 7 \quad 7 \quad 7\frac{1}{2} \quad 8$

Then they drew a line plot to show the data.

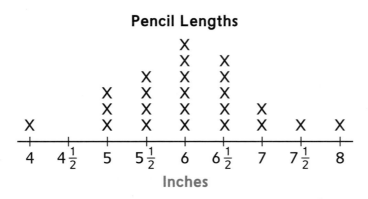

Pencil Lengths

Inches

The line plot shows the measurement data scaled in $\frac{1}{2}$-inch intervals.

Example (continued)

If the children had measured their pencils to the nearest inch, their line plot would look like the one below.

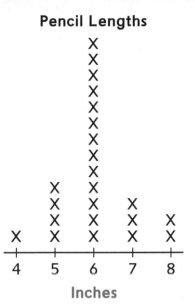

Pencil Lengths

Inches

There are more Xs above each number on this line plot with a 1-inch scale. Using a $\frac{1}{2}$-inch scale allows the children to be more precise about how long their pencils are.

Check Your Understanding

1. Use the Pencil Lengths line plot on page 196 to answer these questions.

   a. How many children's pencils measured 6 inches or more?

   b. How many children had the longest pencil length?

   c. How many more children had 6-inch pencils than 5-inch pencils?

   Check your answers in the Answer Key.

Making Sense of Data

Once you have collected, organized, and represented data, you can look at it to see what you notice. You can tell which number is the smallest (minimum) and which number is the largest (maximum). For example, in the line plots below, the minimum pencil length is 4 inches and the maximum pencil length is 8 inches. The minimum crayon length is $2\frac{1}{4}$ inches and the maximum crayon length is $3\frac{1}{2}$ inches.

Sometimes you can compare two sets of data.

Example

What do you notice when you compare these line plots?

Pencil Lengths

```
                    X
                    X   X
                X   X   X
            X   X   X   X
            X   X   X   X   X
    X       X   X   X   X   X   X   X
    +---+---+---+---+---+---+---+---+
    4  4½   5  5½   6  6½   7  7½   8
                 Inches
```

Crayon Lengths

```
                X
                X
                X   X
            X   X   X
        X   X   X   X
        X   X   X   X               X
    +---+---+---+---+---+---+
    2¼  2½  2¾   3  3¼  3½
              Inches
```

Some of the things you might notice are:

• The pencils were measured to the nearest $\frac{1}{2}$ inch.

  The crayons were measured to the nearest $\frac{1}{4}$ inch.

• The longest pencil is 8 inches. The longest crayon is $3\frac{1}{2}$ inches.

• The shortest pencil is 4 inches. The shortest crayon is $2\frac{1}{4}$ inches.

• The crayons are all shorter than the pencils.

Timekeeping

In the past, people gauged the passage of time by observing nature. They kept track of the seasons by paying attention to migrating birds, falling leaves, and changes in sun, moon, and star positions.

The leaves of many types of trees change color when cooler weather arrives.

Some birds migrate back to their breeding grounds when the weather begins to get warmer.

Constellations, like Orion, appear in different parts of the sky during different seasons.

By observing shadows, we can tell whether it is morning, noon, or evening. In the evening and morning, the sun is closer to the horizon, so shadows are longer. Around noon, the sun is higher in the sky, so shadows are shorter.

We use many clues to figure out what time of day it is. What time of day do you think it is in this photo?

What clues in nature do you use to tell the time or the season?

Tools for Tracking Time

At some point, people figured out how to make tools to track the passage of time. Early tools relied on the movements of celestial bodies like the sun, moon, and stars.

These huge stones in England, known as Stonehenge, were set up nearly 5,000 years ago. Stonehenge can indicate when summer and winter begin.

As early as 3,500 BCE, the Egyptians and Sumerians built shadow clocks. As the sun moves across the sky, the length and direction of the shadow give a rough idea of the time of day.

Advancements in shadow clocks led to sundials. Sundials can be very accurate, but they have limitations. For example, they do not work on very cloudy days.

Pendulums and Pendulum Clocks

An important advancement in clock making came about with the invention of the pendulum clock.

A *pendulum* is a weight hanging from a string or stick that can swing back and forth.

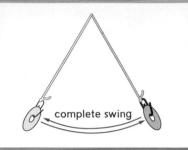

complete swing

In the 1580s, the Italian scientist Galileo Galilei made an important discovery. He noticed that it took the same amount of time for a chandelier to complete one swing no matter how wide the swing.

About 75 years later, Dutch scientist Christiaan Huygens used Galileo's discovery to make the first pendulum clock. These clocks use a swinging pendulum to keep the time.

This grandfather clock uses a pendulum to measure time. As Galileo discovered, each swing takes the same amount of time.

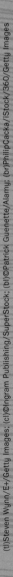

Modern Clocks

Clocks have become more accurate and more versatile over time.

In the 1920s, timekeeping took a giant leap forward as inventors took advantage of a property of quartz crystals. Running electricity through these crystals causes them to vibrate at a constant rate, like the constant rate of a pendulum swing.

Today, quartz clocks are built into many watches, calculators, and personal computers. They are popular timekeepers because they are inexpensive and accurate.

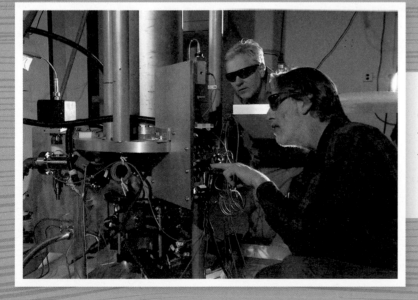

In 1955, scientists built a clock that is more accurate than a quartz clock. It is called the cesium atomic clock. It is accurate to 1 second every 1,000,000 years.

There are many places where you can find the time of day.

Many clocks today are based on time from atomic clocks. Places like the National Institute of Standards and Technology have atomic clocks and transmit time signals through radio waves to smart clocks. They run computer servers, which makers of computers, smart phones, and other devices connect to through the Internet so the devices can keep accurate time.

Appliances like this oven display the time.

There are probably clocks like this one in your school.

Cities have clocks like Big Ben in London.

Where do you look when you need to know what time it is?

Geometry

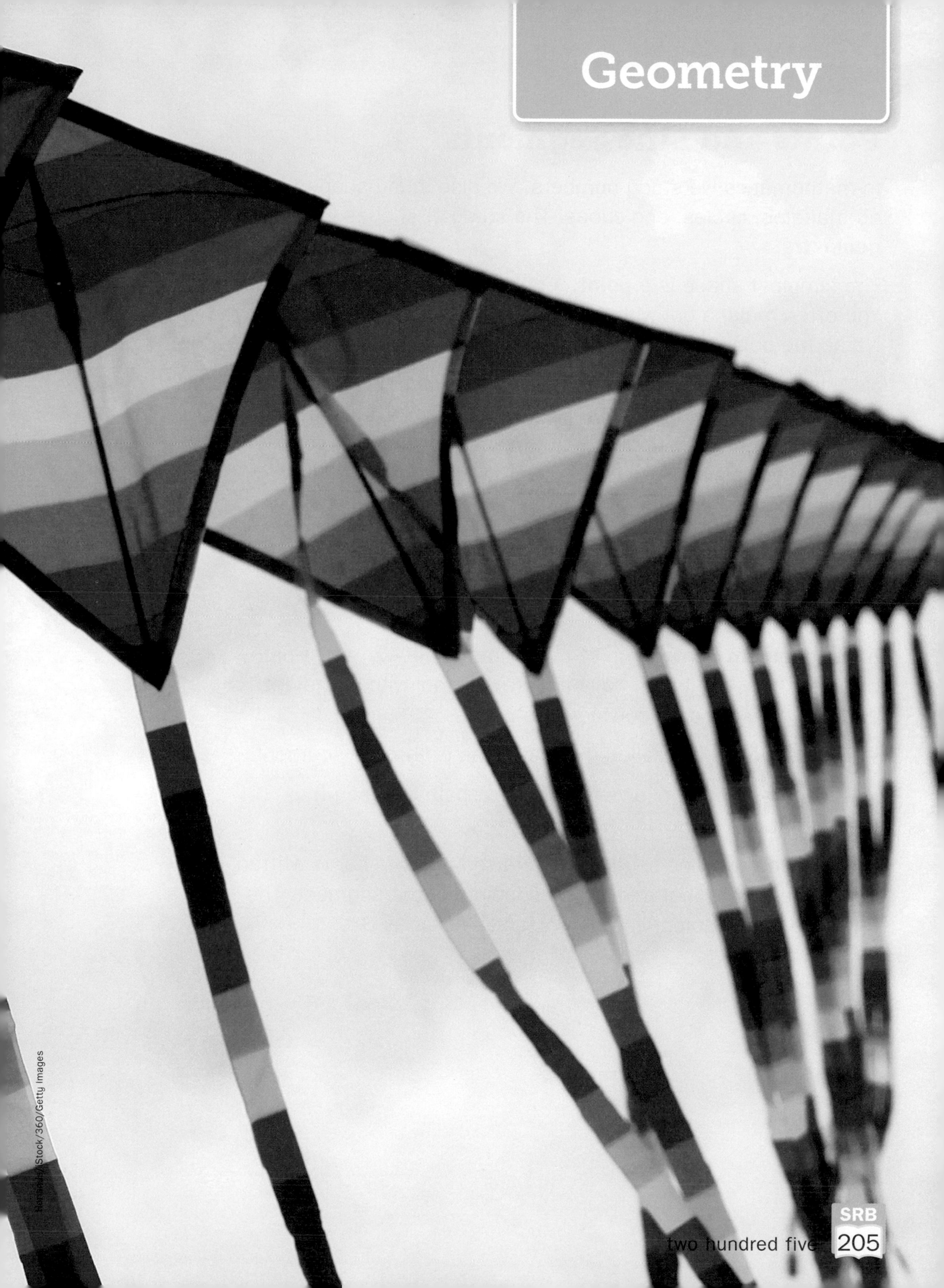

Points and Line Segments

In mathematics we study numbers. We also study shapes such as triangles, circles, and cubes. The study of shapes is called **geometry.**

The simplest shape is a **point.** A point is a location in space. You often make a dot with a pencil to show where a point is. Name the point with a capital letter.

Here is a picture of three points. The letter names make it easy to talk about the points. For example, point *A* is closer to point *B* than it is to point *P*. And point *B* is closer to point *A* than it is to point *P*.

A   B
•   •

•
P

A **line segment** is made up of two points and the straight path of points between them. You can use any tool with a straight edge to draw the path between two points.

- The two points are called the **endpoints** of the line segment.
- The line segment follows the shortest path between the endpoints.

The symbol for a line segment is a raised bar. The bar is written above the letters that name the endpoints of the segment. The following line segment can be written as $\overline{AB}$ or as $\overline{BA}$.

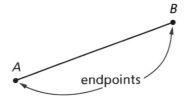

endpoints

Rays and Lines

A **ray** is a straight path of points that starts at *one* endpoint and goes on forever in *one* direction. You can draw a line segment with one arrowhead to stand for a ray.

Point *R* is the **endpoint** of this ray. The symbol for a ray is a raised arrow pointing to the right. The ray shown here can be written $\overrightarrow{RA}$. The endpoint *R* is listed first. The second letter names some other point on the ray.

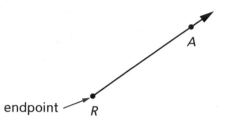

endpoint — *R*

A **line** is a straight path of points that goes on forever in *both* directions. You can draw a line segment with two arrowheads to stand for a line. The symbol for a line is a raised bar with two arrowheads.

You can name a line by listing two points on the line. Then write the symbol for a line above the letters. The line here is written as $\overleftrightarrow{FE}$ or as $\overleftrightarrow{EF}$.

Imagine these are straight lines that go on forever in both directions.

E   *F*

Example

Write all the names for this line.

Points *D*, *O*, and *G* are all on the line. Use any two points to write the name of the line.

$\overleftrightarrow{DO}$ or $\overleftrightarrow{OD}$ or $\overleftrightarrow{DG}$ or $\overleftrightarrow{GD}$ or $\overleftrightarrow{OG}$ or $\overleftrightarrow{GO}$

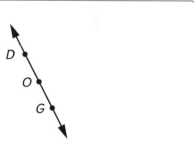

Angles

An **angle** is formed by two rays, two line segments, or a ray and a line segment that share the same endpoint.

angle formed by 2 rays

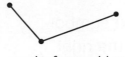
angle formed by 2 segments

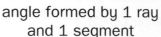
angle formed by 1 ray and 1 segment

The endpoint where the rays or line segments meet is called the **vertex** of the angle. The rays or line segments are called the **sides** of the angle.

∠ is the symbol for an angle.
This is angle *T*, or ∠*T*.

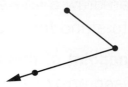

an angle on a soccer field

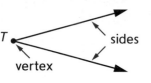

Angles can be found in many shapes. For example, this triangle has three angles: ∠*K*, ∠*J*, and ∠*E*.

Angle Measures

Angles are measured in degrees. A **right angle** measures 90° (90 degrees). Its sides form a square corner. A small corner symbol inside the angle shows that it is a right angle.

Examples

A curved red arrow in each picture shows which angle opening is measured.

Measure of ∠*A* is 60°.

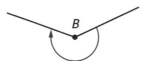
Measure of ∠*B* is 225°.

Measure of ∠*C* is 300°.

Parallel Lines and Segments

Parallel lines are lines that never meet or cross. Imagine a straight railroad track that goes on forever. The two rails are parallel. The rails never meet or cross, and they are always the same distance apart.

Parallel line segments are segments that are parts of lines that are parallel. A section of a straight railroad track has two rail segments that are parallel. Some polygons have parallel line segments. Two pairs of sides in rectangle *CMAJ* are parallel line segments, as shown to the right.

The symbol for *parallel* is a pair of vertical lines ∥.

If lines or segments cross or meet each other, they **intersect**.

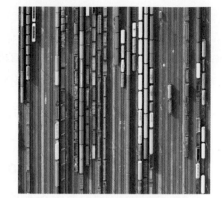

aerial photo of a railroad yard

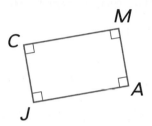

Line segments *CJ* and *MA* are parallel. Line segments *CM* and *JA* are parallel.

intersecting roads

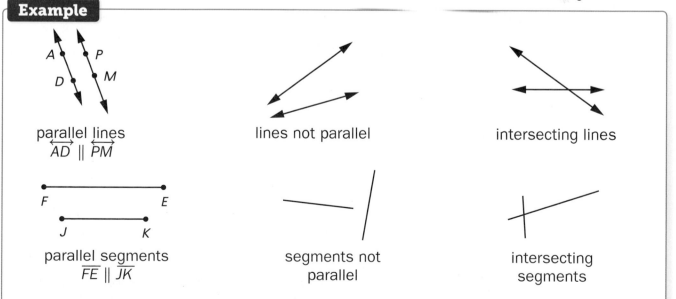

Example

parallel lines
$\overleftrightarrow{AD} \parallel \overleftrightarrow{PM}$

lines not parallel

intersecting lines

parallel segments
$\overline{FE} \parallel \overline{JK}$

segments not parallel

intersecting segments

Polygons

A **polygon** is a flat, 2-dimensional figure made up of three or more line segments called **sides.**

- The sides of a polygon are connected end to end and make one closed path.
- The sides of a polygon do not cross.

Each endpoint where sides meet is called a **vertex.** The plural of the word *vertex* is **vertices.**

Figures That Are Polygons

4 sides, 4 vertices

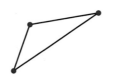
3 sides, 3 vertices

7 sides, 7 vertices

Figures That Are NOT Polygons

All sides of a polygon must be line segments. Curved lines are not line segments.

The sides of a polygon must form a closed path.

A polygon must have at least three sides.

The sides of a polygon must not cross.

Polygons are named based on their number of sides.
The prefix in a polygon's name tells the number of sides it has.

triangle

quadrilateral

pentagon

hexagon

heptagon

octagon

nonagon

decagon

Prefixes	
tri-	3
quad-	4
penta-	5
hexa-	6
hepta-	7
octa-	8
nona-	9
deca-	10
dodeca-	12

Did You Know?

The flag of Nepal is the only flag in the world with five sides. All other flags have four sides. The flag of Switzerland has a white cross with an edge that is a dodecagon (12 sides).

flag of Nepal

flag of Switzerland

Check Your Understanding

1. Name the polygon with the following number of sides.

   a. 6 sides   b. 4 sides   c. 10 sides
   d. 8 sides   e. 12 sides

2. Draw a pentagon whose sides are not all the same length.

Check your answers in the Answer Key.

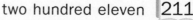

A **regular polygon** is a polygon whose sides all have the same length and whose angles are all the same size.

The hexagon shown here is a regular hexagon.
All six sides have the same length.
All six angles are the same size.

Some regular polygons have special names.

A regular triangle is called an **equilateral triangle.** It has three sides that are all the same length. It has three angles that each measure 60°.

A **square** is a regular quadrilateral (quadrangle). It has four sides that are the same length. It has four angles that each measure 90°.

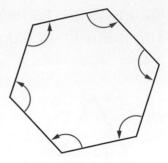

regular hexagon

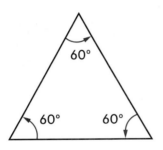

equilateral triangle

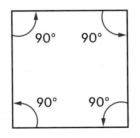

square

a design using
equilateral triangles

laying a square
tile floor

The Pentagon in Washington, D.C. has a ground area of 1,263,240 square feet. It has a larger ground area than any other office building in the world. The outside walls of the Pentagon have the shape of a regular pentagon. The inside walls of the Pentagon also have the shape of a regular pentagon. The five angles of a regular pentagon each measure 108°.

A stop sign has the shape of a regular octagon. In the United States, no other traffic sign has this shape. This makes it easier for drivers to identify the sign from a distance, in the dark, and even when it is covered with snow. It has eight equal-length sides. It has eight angles that each measure 135°.

Triangles

Triangles have fewer sides and angles than any other polygon. The prefix *tri-* means *three*. All triangles have three sides, three vertices, and three angles.

Example

Name the parts of the triangle below.

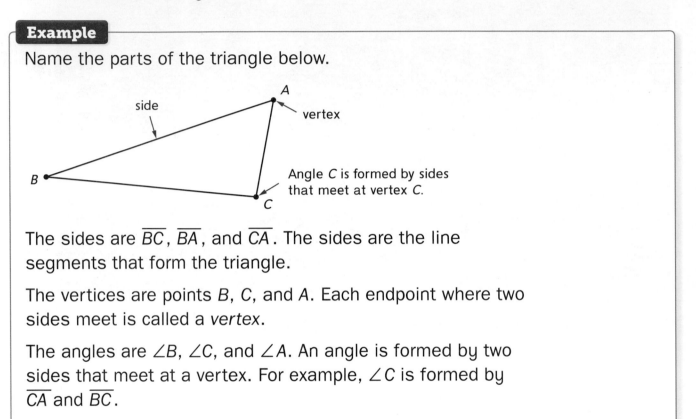

The sides are $\overline{BC}$, $\overline{BA}$, and $\overline{CA}$. The sides are the line segments that form the triangle.

The vertices are points *B*, *C*, and *A*. Each endpoint where two sides meet is called a *vertex*.

The angles are ∠*B*, ∠*C*, and ∠*A*. An angle is formed by two sides that meet at a vertex. For example, ∠*C* is formed by $\overline{CA}$ and $\overline{BC}$.

The symbol for a triangle is △. Triangles have three-letter names. You name a triangle by listing the letters for each vertex in order. The triangle in the example above has six possible names:

△*BCA*, △*BAC*, △*CAB*, △*CBA*, △*ABC*, and △*ACB*.

Which blocks are shaped like triangles?

supermimicry/iStock/360/Getty Images

Triangles have many different sizes and shapes. Special types of triangles have been given names.

Equilateral Triangles

An **equilateral triangle** is a triangle with all three sides the same length. It has three angles that each measure 60°. All equilateral triangles have the same shape.

Right Triangles

A **right triangle** is a triangle with one right angle (square corner). Right triangles have many different shapes.

Other triangles are shown below. None of these is an equilateral triangle. None is a right triangle.

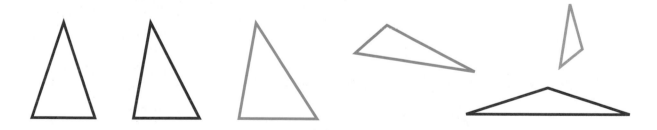

Quadrilaterals

A **quadrilateral** is a polygon with four sides. The prefix *quad-* means *four*. All quadrilaterals have four sides and four vertices. Another name for *quadrilateral* is **quadrangle.**

Example

Name the parts of the quadrilateral.
The sides are $\overline{RS}$, $\overline{ST}$, $\overline{TU}$, and $\overline{UR}$.
The vertices are R, S, T, and U.
The angles are ∠R, ∠S, ∠T, and ∠U.

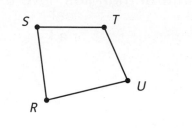

Some quadrilaterals have *at least* one pair of parallel sides. These quadrilaterals are called **trapezoids.**

Reminder: The sides of a shape are parallel if they are always the same distance apart. If you extend the line segments in both directions, the lines will always be the same distance apart.

Note Different mathematicians may use different definitions. In *Everyday Mathematics*, a **trapezoid** is defined as a quadrilateral that has *at least* one pair of parallel sides. With this definition, parallelograms are also trapezoids. Other mathematicians define a trapezoid as a quadrilateral with *exactly* one pair of parallel sides. With this definition, parallelograms are not trapezoids since parallelograms have more than one pair of parallel sides.

Figures That Are Trapezoids

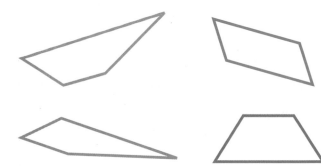

Each figure has *at least* one pair of parallel sides.

Figures That Are NOT Trapezoids

 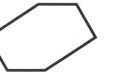 

no parallel sides   not a quadrilateral   no parallel sides

Quadrilaterals: Examples and Definitions

Many special types of quadrilaterals have been given names.

Trapezoids are quadrilaterals with at least one pair of parallel sides.	
Kites are quadrilaterals whose four sides are in two pairs of equal length. The equal-length sides are next to each other. If all four sides of a kite have the same length, the kite is also a rhombus.	
Parallelograms are quadrilaterals that have two pairs of parallel sides.	
Rectangles are parallelograms that have four right angles. The sides do not have to be the same length.	
Rhombuses are parallelograms with four sides that have the same length.	
Squares are parallelograms with four right angles and four sides that have the same length.	

Check Your Understanding

1. Look at the figures on page 216 that are trapezoids. Which shapes have one pair of parallel sides? Which shapes have two pairs of parallel sides?

2. All squares are also rectangles. Explain why.

Check your answers in the Answer Key.

The Pattern-Block Template

There are 13 geometric figures on the Pattern-Block Template. Six of the figures are the same size and shape as the actual pattern blocks.

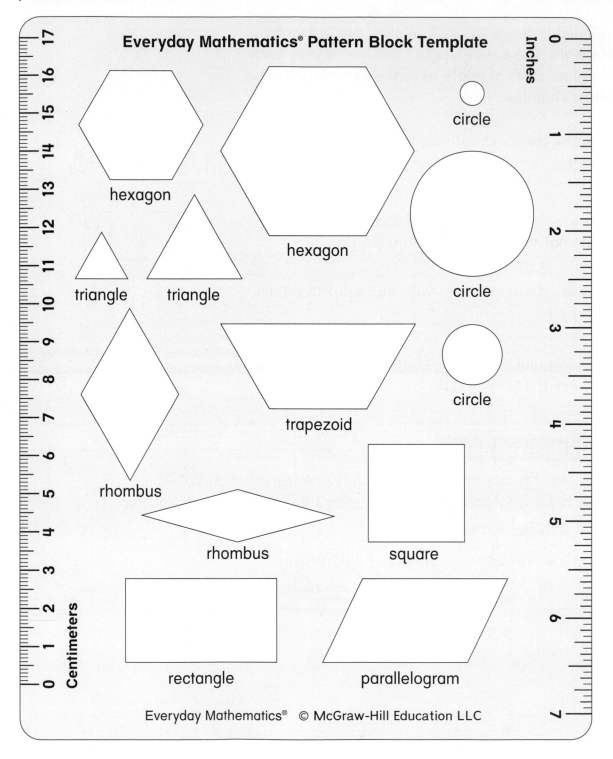

Everyday Mathematics® Pattern Block Template

hexagon

hexagon

circle

circle

circle

triangle

triangle

rhombus

trapezoid

rhombus

square

rectangle

parallelogram

Inches

Centimeters

Everyday Mathematics®   © McGraw-Hill Education LLC

Solids

Triangles, quadrilaterals, and circles are flat shapes. They take up a certain amount of area, but they do not take up space. They are flat, **2-dimensional** figures that you can draw on a sheet of paper, but you can't hold in your hand.

Solid objects that take up space are things such as boxes, books, and chairs. They are **3-dimensional** objects. You can hold small 3-dimensional shapes in your hand. Some solids, such as rocks and animals, do not have regular shapes. Other solids have shapes that are easy to describe using geometric words.

The **surfaces** on the outside of a solid may be flat or curved or both. A **flat surface** of a solid is called a **face.**

The surfaces of a solid meet one another. They form curves or line segments. These curves or line segments are the **edges** of the solid.

The corner of a solid is a **vertex.**

A **cube** has 6 faces, 12 edges, and 8 vertices. All the faces are squares.

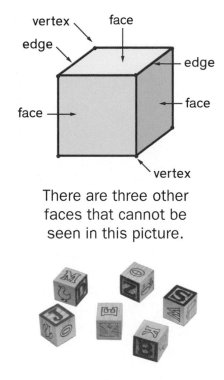

There are three other faces that cannot be seen in this picture.

These blocks are shaped like cubes.

Mark Steinmetz

A **cylinder** has two faces and one curved surface.

A cylinder has two edges and zero vertices.

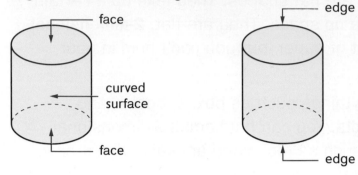

A **cone** has one face and one curved surface.

A cone has one edge and one vertex.

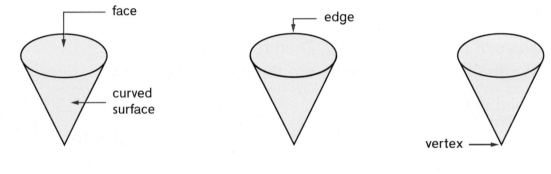

The table top and base are short cylinders with large faces. The center pole is a tall cylinder with small faces.

This ice cream cone is a good example of a cone. But keep in mind that a cone is closed. It has one face that acts like a cover.

(l)Maksym Bondarchuk/iStock/360/Getty Images, (r)MADDRAT/iStock/Getty Images Plus

Geometry in Nature

Many interesting shapes in nature result from the way things grow or form. You can see some of these shapes with the naked eye. You can see others with special equipment, such as magnifying lenses, microscopes, or telescopes.

If you look at one of your fingertips with a powerful magnifying lens, you will see many curves. The curves of each unique fingerprint provide traction so you can lift things.

Because they are so far away, the stars of the Big Dipper look like tiny points of light when viewed from Earth.

Polygons, Circles, and Spheres

If you imagine a line from tip to tip of each arm of this sea star, you will make a pentagon.

Honeycombs look like connected hexagons. These hexagons fit together with no space in between. They make a great place for bees to store honey.

Look closely at the shapes in this spider web. Do you see shapes like quadrilaterals?

The head of a dandelion that has gone to seed is in the shape of a sphere.

The circular leaves of the *Victoria amazonica* water lily can grow so large that two metersticks would fit across a leaf from one side to the other.

If you cut an onion in half, you will see what look like concentric circles, or circles within circles.

Spirals

This is a drawing of a spiral. Spirals can be found in nature.

This photograph was taken above the clouds of Hurricane Katrina in 2005. Do you see the spirals?

The Meller's chameleon coils its tail into a spiral.

Spirals can also be found in many plants like *Aloe polyphylla* (spiral aloe).

Three-Dimensional Shapes and Solids

When salt crystals are magnified using a high-powered microscope, you can see that they are rectangular solids.

These quartz crystals are geometric solids. You can see some of their many faces.

The rock columns of the Giant's Causeway in Ireland were shaped during volcanic activity. Some of the columns have a hexagonal shape.

The cone-shaped funnel cloud of a tornado descends to the ground.

Symmetry

Look at this picture of a butterfly. A dashed line is drawn through it. The line divides the butterfly into two parts. Both parts look alike but are facing in opposite directions. The figure is *symmetric*, and the dashed line is called a *line of symmetry.*

The left and right sides of the stag beetle are symmetrical.

The two halves of this artichoke are symmetrical.

Magnifying snowflakes shows that they are symmetrical. What shapes do you see in this snowflake?

This is a close-up of an orchid flower. Do you see the symmetry between one side and the other?

Look around you. Where do you see geometry in nature?

(t)blackwaterimages/iStock/360/Getty Images; (tr)Siede Preis/Getty Images; (c)Olga Katrychenko New Zealand/Moment Open/Getty Images; (cr) ©Foodcollection; (br)Melba Photo Agency/Alamy

Games

Throughout the year, you will play games that help you practice important math skills. Playing mathematics games gives you a chance to practice math skills in a different way. In this section of your *Student Reference Book,* you will find the directions for many games. We hope that you will play often and have fun.

Materials

The materials for each game are different and may include cards, dice, coins, counters, and calculators. For some games you will have to make a gameboard or a score sheet. These instructions are included with the game directions. For other games, your teacher will provide gameboards and card decks.

Number Cards. You need a deck of number cards for many of the games. You can use an Everything Math Deck, a deck of regular playing cards, or make your own deck out of index cards. An Everything Math Deck is part of the eToolkit.

An Everything Math Deck includes 54 cards. There are 4 cards each for the numbers 0–10. And there is 1 card for each of the numbers 11–20.

You can also use a deck of regular playing cards after making a few changes. A deck of playing cards includes 54 cards (52 regular cards, plus 2 jokers). To create a deck of number cards, use a permanent marker to mark the cards in the following ways:

- Mark each of the 4 aces with the number 1.

- Mark each of the 4 queens with the number 0.

- Mark the 4 jacks and 4 kings with the numbers 11 through 18.

- Mark the 2 jokers with the numbers 19 and 20.

Game	Skill	Page
The Area and Perimeter Game	Calculating area and perimeter of rectangles	230–231
Array Bingo	Modeling multiplication with arrays	232–233
Baseball Multiplication	Practicing multiplication facts	234–236
Beat the Calculator	Practicing multiplication or division facts	237
Division Arrays	Modeling division with and without remainders	238–239
Factor Bingo	Recognizing products of given factors	240–241
Finding Factors	Identifying factors and products of basic multiplication facts	242
Fraction Memory	Recognizing equivalent fractions	243–244
Fraction Number-Line Squeeze	Locating fractions on a number line	245
Fraction Top-It	Comparing fractions	246–247
Multiplication Draw	Practicing multiplication facts	248
Name That Number	Finding equivalent names for numbers	249–250
Number-Grid Difference	Using a number grid to find the difference between 2-digit numbers	251
Product Pile-Up	Practicing multiplication facts	252
Roll to 1,000	Adding multiples of 10	253–254
Salute!	Practicing multiplication and division facts	255
Shuffle to 100	Estimating sums and making combinations close to 100	256–257
Spin and Round	Rounding numbers to the nearest 10 or 100	258–259
Top-It Games	Practicing basic facts	260–261
What's My Polygon Rule?	Understanding attributes of polygons	262

The Area and Perimeter Game

Materials ☐ 1 *The Area and Perimeter Game* Action Deck
(*Math Journal 1*, Activity Sheet 13)

☐ 1 *The Area and Perimeter Game* Deck A
(*Math Journal 1*, Activity Sheet 14)

☐ 1 *The Area and Perimeter Game* Record Sheet for each player
(*Math Masters*, p. G16)

☐ calculator (optional)

Players 2

Skill Calculating area and perimeter

Object of the Game To score more points by finding the areas and
perimeters of rectangles.

Directions

1. Shuffle *The Area and Perimeter Game* Action Deck and place it word-side
down on the table. Shuffle *The Area and Perimeter Game* Deck A and
place it picture-side down next to the action cards.

2. Players take turns. When it is your turn, draw 1 card from each deck
and place the cards faceup on the table.

   - If an area (A) card is drawn, the player finds the area.

   - If a perimeter (P) card is drawn, the player finds the perimeter.

   - If a "Player's Choice" card is drawn, the *player* may choose to find
   either the area or perimeter. See the example on the next page.

   - If a "Partner's Choice" card is drawn, your *partner* chooses whether
   you will find the area or perimeter.

3. Record your turns on your record sheet. During your turn, record the
rectangle card number and write A (area) or P (perimeter). Show how you
found the area or perimeter. The solution is your score for the round.

4. The player with the higher total score at the end of 6 rounds is the
winner. Players may use a calculator to find their total scores.

Example

You draw the two cards to the right. You may choose to calculate the area or the perimeter. Before you answer, figure out both the area and the perimeter in your head.

A or P

Player's Choice

You count by 4s to find the area: 4, 8.
Area = 8 square units

and

You add the side lengths of the rectangle to find the perimeter.
2 + 4 + 2 + 4 = 12
Perimeter = 12 units

This is a 2-by-4 rectangle.

3

You choose to record the perimeter, because that will earn you more points.

Round	Card Number	A (area) or P (perimeter)	Show how you found the area or perimeter using words, drawings, and/or a number sentence.	Score
Example:	3	P	*I counted the boxes along the long and short sides of the rectangle and added them:* 2 + 4 + 2 + 4 = 12 *units.*	12

Variation

Play using *The Area and Perimeter Game* Deck B (*Math Journal 1*, Activity Sheet 15), which shows blank rectangles and two side lengths.

Array Bingo

Materials ☐ 1 set of *Array Bingo* Cards for each player (*Math Masters*, p. G8)

☐ number cards 1–20 (1 of each)

Players 2 or 3

Skill Modeling multiplication with arrays

Object of the Game To have a row, column, or diagonal of cards facedown.

Directions

1 Each player arranges his or her *Array Bingo* Cards faceup in a 4-by-4 array.

2 Shuffle the number cards. Place them number-side down.

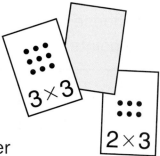

3 Players take turns. When it is your turn, draw a number card. Look for any one of your array cards with that number of dots and turn it facedown. If there is no matching array card, your turn ends. Place your number card in a discard pile.

4 The first player to turn a card facedown so that a row, column, or diagonal of cards is all facedown calls out "Bingo!" See example on the next page.

5 If all the number cards are used before someone wins, shuffle the deck and continue playing.

Example

Abby draws the number card 4 when it is her turn.

She can turn over the card with the 2 × 2 array.

After Abby turns over the 2 × 2 array card, there are 4 facedown cards in a row. She calls out, "Bingo!"

Baseball Multiplication

Materials ☐ 1 *Baseball Multiplication* (with 10-Sided Dice) Game Mat (*Math Masters*, p. G17)

☐ 2 ten-sided dice (labeled 1–10)

☐ 4 counters

Players 2 teams of one or more players each

Skill Practicing multiplication facts

Object of the Game To score more runs in a 3-inning game.

Directions

The rules are similar to the rules of baseball, but this game lasts only 3 innings. In each inning, each team bats until it makes 3 outs. Each team rolls a die to decide who bats or pitches first (the highest roll gets to choose). The team with more runs when the game is over wins.

Pitching and Batting: Members of the team not at bat take turns "pitching." They roll the dice to get two factors. Players on the "batting" team take turns multiplying the two factors and saying the product.

The pitching team checks the product. An incorrect answer is a strike, and another pitch (dice roll) is thrown. Three strikes make an out.

Hits and Runs: If the batting team's answer is correct, the batter checks the Scoring Chart on the game mat. If the chart shows a hit, the batter moves a counter to a base as shown in the Scoring Chart. Runners already on base are moved ahead of the batter by the same number of bases. A run is scored every time a runner reaches home plate.

Keeping Score: For each inning, keep a tally of runs scored and outs made. Use the Runs-and-Outs Tally on the game mat. At the end of the inning, record the number of runs on the Scoreboard.

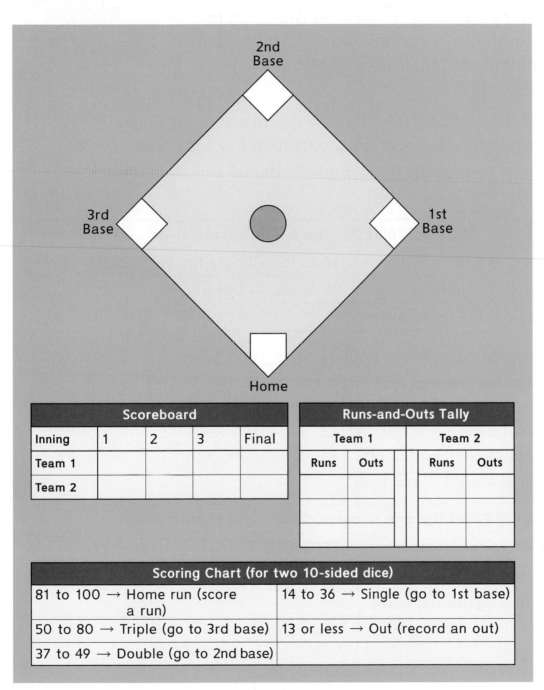

Scoreboard				
Inning	1	2	3	Final
Team 1				
Team 2				

Runs-and-Outs Tally			
Team 1		Team 2	
Runs	Outs	Runs	Outs

Scoring Chart (for two 10-sided dice)	
81 to 100 → Home run (score a run)	14 to 36 → Single (go to 1st base)
50 to 80 → Triple (go to 3rd base)	13 or less → Out (record an out)
37 to 49 → Double (go to 2nd base)	

Baseball Multiplication (with 10-Sided Dice) Game Mat

Variations

Baseball Multiplication (with Number Cards 1–10): Use number cards 1–10 (4 of each) instead of 10-sided dice. Players shuffle the cards and place the deck number-side down. Players on the "pitching" team draw two cards to generate two factors. Players on the "batting" team take turns finding the product. Teams play and keep score on *Math Masters*, page G17.

Baseball Multiplication (with 6-Sided Dice): Players on the "pitching" team roll two 6-sided dice to generate two factors. Players on the "batting" team take turns finding the product. Teams play and keep score on *Math Masters*, page G18.

Scoring Chart (for two 6-sided dice)	
36 → Home run (score a run)	6 to 15 → Single (go to 1st base)
25 to 35 → Triple (go to 3rd base)	5 or less → Out (record an out)
16 to 24 → Double (go to 2nd base)	

Baseball Multiplication (with Tens): Players on the "pitching" team roll one 10-sided die and multiply the number rolled by 10. For example, if a 6 is rolled, it is 60. Players on the "batting" team turn over a number card (1–10) and take turns finding the product of the two factors. Teams play and keep score on *Math Masters*, page G19.

Scoring Chart (for one 10-sided die and one number card)	
810 to 1,000 → Home run (score a run)	140 to 300 → Single (go to 1st base)
500 to 800 → Triple (go to 3rd base)	130 or less → Out (record an out)
310 to 490 → Double (go to 2nd base)	

Beat the Calculator

Materials ☐ 1 set of Fact Triangles

☐ 1 calculator

☐ *Beat the Calculator* Triangle (*Math Masters*, p. G20; optional)

Players 3

Skill Practicing multiplication or division facts

Object of the Game To multiply numbers without a calculator faster than a player using one.

Directions

1. The three players each choose a role: *Caller*, *Calculator*, or *Brain*.

2. Place the Fact Triangles in a pile on the table. The Caller draws one Fact Triangle from anywhere in the pile, covering the product (the number by the dot). The Caller asks for the product of the numbers.

3. The Calculator solves the problem *with* a calculator. The Brain solves it *without* a calculator. The Caller says who answered faster.

4. The Caller continues to draw Fact Triangles from the deck and ask for the product of the numbers. Players trade roles every 10 turns.

Example

The Caller draws this Fact Triangle, covers the number 70, and says, "10 times 7." The Brain and the Calculator each solve the problem. The Caller says who answered faster.

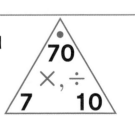

Variations

- Instead of covering the product, the Caller covers one of the factors. The Caller says the division problem and asks for the quotient.

- Play with number cards 0–10 (4 of each) instead of fact triangles. The Caller draws 2 cards and asks for the product of the numbers.

- Play with number cards 0–10 (4 of each). The Caller draws 2 cards and asks for the sum of the numbers.

Division Arrays

Materials
☐ number cards 6–18 (1 of each)

☐ 1 six-sided die

☐ 18 counters

☐ 1 *Division Arrays* Record Sheet for each player (*Math Masters*, p. G9)

Players 2 to 4

Skill Modeling division with and without remainders

Object of the Game To have the highest total score.

Directions

1 Shuffle the cards. Place the deck number-side down on the table.

2 Players take turns. When it is your turn, draw a card and take the number of counters shown on the card. You will use the counters to make an array.

• Roll the die. The number on the die is the number of equal rows you must have in your array.

• Make an array with the counters.

• Your score is the number of counters in one row. If there are no leftover counters, your score is double the number of counters in one row. See the example on the next page.

3 Players keep track of their scores. The player with the highest total score at the end of 5 rounds wins.

Example

Number card	Die	Array formed	Number model	Leftovers?	Score
10	2	:::::	10 ÷ 2 = 5	No	10
9	2	::::	9 ÷ 2 → 4 R1	• Yes	4
14	3	:::	14 ÷ 3 → 4 R2	: Yes	4
11	6	⋮	11 ÷ 6 → 1 R5	⋮ Yes	1

Factor Bingo

Materials	☐ number cards 2–10 (4 of each)
	☐ 1 *Factor Bingo* Game Mat for each player (*Math Masters*, p. G24)
	☐ 12 counters per player
Players	2 to 4
Skill	Recognizing products of given factors
Object of the Game	To get 5 counters in a row, column, or diagonal; or to get 12 counters anywhere on the game mat.

Directions

1 Fill in your game mat. Choose 25 different numbers from 2 to 90.

2 Write each number you choose in exactly one square on your game mat grid. The numbers should not all be in order, so be sure to mix them up as you write them on the grid. To help you keep track of the numbers you use, circle them in the list below the game mat.

3 Shuffle the cards and place them number-side down on the table. Any player can turn over the top card. This card is the *factor*.

4 Players check their grids for a product that has the card number as a factor. Players who find a product cover it with a counter. A player may place only one counter on the grid for each card.

5 Turn over the next card and repeat Step 4. The first player to get 5 counters in a row, column, or diagonal calls out "Bingo!" and wins. A player may also win by getting 12 counters anywhere on the game mat.

6 If all of the cards are used before someone wins, shuffle the cards again and continue playing.

> **Example**
>
> A 5 card is turned over. The number 5 is the *factor*. Players may place a counter on one number that is a product of 5, such as 5, 10, 15, or 20.

Factor Bingo Game Mat

Choose any 25 *different* numbers from the numbers 2 through 90. Write each number you choose in exactly 1 square on your game mat page. To help you keep track of the numbers you use, circle them in the list on your game mat page.

2	3	4	5	6	7	8	9	10	
11	12	13	14	15	16	17	18	19	20
21	22	23	24	25	26	27	28	29	30
31	32	33	34	35	36	37	38	39	40
41	42	43	44	45	46	47	48	49	50
51	52	53	54	55	56	57	58	59	60
61	62	63	64	65	66	67	68	69	70
71	72	73	74	75	76	77	78	79	80
81	82	83	84	85	86	87	88	89	90

Factor Bingo **Game Mat**

Variation

Speed Factor Bingo: The directions are similar to *Factor Bingo* except for these changes:

- This is a one-player game. The object of the game is to be able to call "Bingo!" as quickly as possible.

- Use number cards 2–10 (only 1 of each).

- When you turn over the top card to get the factor, you may place counters on *all* products of the factor on your game mat. Continue until you can call "Bingo!" or until all of the cards are used.

- If you can call "Bingo!" count the number of cards you turned over. This is your score.

- If you cannot call "Bingo!" before all the cards are turned over, your score is 10.

- Make a new game mat or revise the one you have to try to get a lower score. Play again.

Finding Factors

Materials
☐ 1 *Finding Factors* Gameboard (*Math Masters*, p. G23)

☐ 2 counters (different colors)

☐ 2 crayons (different colors)

Players 2

Skill Identifying factors and products of basic multiplication facts

Object of the Game To win 5 squares in a row, column, or diagonal.

Directions

1 Each player chooses a counter and a crayon.

2 Player 1 places his or her counter on one of the factors in the factor strip at the bottom of the gameboard.

3 Player 2 places his or her counter on one of the factors in the factor strip. (Two counters can cover the same factor.)

4 Player 2 multiplies the two factors to find the product and wins the square by naming the product. He or she colors in the square at the top of the gameboard.

5 Player 1 moves **one** of the counters to a new factor on the factor strip and finds the product of the two covered factors. If the product has not been colored, he or she wins the square and colors it in.

6 Play continues until one player has won 5 squares in a row, column, or diagonal.

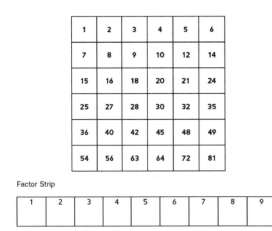

Factor Strip

Finding Factors Gameboard

Fraction Memory

Materials ☐ 1 set of fraction cards (*Math Journal 2*, Activity Sheets 16–18)

Players 2

Skill Recognizing equivalent fractions

Object of the Game To collect the most pairs of equivalent fraction cards.

Directions

Before playing, remove the following cards from the deck:
$\frac{0}{1}, \frac{1}{6}, \frac{5}{6}, \frac{1}{8}, \frac{3}{8}, \frac{5}{8}$, and $\frac{7}{8}$.

1. Create a 4-by-5 array of fraction cards with the picture side facing down.

2. Players take turns. When it is your turn:

   - Turn over any 2 cards so the other player can see the picture side of the cards.

   - Determine whether the fractions are equivalent using the shading on the circles, the location of the points on the number lines, or other reasoning.

   - If the fractions are equivalent, you keep the pair of cards.

   - If the fractions are not equivalent, turn the cards back over in their places.

   With either outcome, your turn is over. See the example on the next page.

4. The game is over when all fraction cards have been matched and removed from the array. The player with the most matches wins.

Example

Damien turns over the $\frac{1}{2}$ card and the $\frac{2}{4}$ card. He sees that the shaded areas of the circles on the cards are equal. These are equivalent fractions. He gets to keep the pair of cards. His turn is finished.

Samantha turns over the $\frac{3}{4}$ card and the $\frac{4}{6}$ card. She sees that the points on the number lines are different distances from 0. These are not equivalent fractions. She turns the cards back over in their places in the array. Her turn is finished.

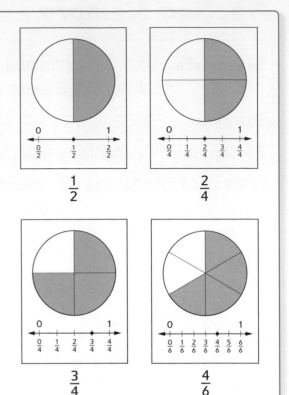

Variation

Rules and play are the same as the regular game, except:

• Use the fraction cards from *Math Journal 2*, Activity Sheets 16–21. Remove the following cards: $\frac{0}{1}$, $\frac{0}{2}$, $\frac{1}{8}$, $\frac{3}{8}$, $\frac{5}{8}$, and $\frac{7}{8}$.

• Create a 6-by-8 array of cards.

Fraction Number-Line Squeeze

Materials ☐ 1 number line (*Math Masters*, p. TA50)

☐ 2 counters

☐ scissors

☐ tape

Players 2 or more

Skill Locating fractions on a number line

Object of the Game To locate the mystery fraction.

Directions

1 Cut out the number line sections from *Math Masters*, page TA50. Tape the sections together to make a number line starting at 0.

2 Choose a denominator: 2, 3, 4, 6, or 8. Starting at 0, label your number line with fractions with that denominator. Label whole numbers with a fraction and a whole number.

3 Place a counter at each end of your number line.

4 One player, the leader, silently chooses a mystery fraction on the number line. Other players take turns guessing the fraction.

5 The leader says whether the guess is *greater than* or *less than* the mystery fraction.

- If the guess is *less than* the mystery fraction, the leader moves the left counter to cover the guessed number.

- If the guess is *greater than* the mystery fraction, the leader moves the right counter to cover the guessed number.

6 Continue playing until one player guesses the mystery fraction or the fraction is "squeezed" between the two counters.

7 The player who guesses the mystery fraction is the next leader.

Variation

Choose any denominator between 2 and 12.

Fraction Top-It

Materials	☐ 1 set of fraction cards (*Math Journal 2*, Activity Sheets 16–18)
Players	2
Skill	Comparing fractions
Object of the Game	To collect more cards.

Directions

1 Shuffle the fraction cards and place the deck picture-side down on the table.

2 Each player turns over a card from the top of the deck.

3 Players take turns. When it is your turn:

- Compare the two fractions using the picture side of the cards. The player with the larger fraction takes both cards.

- If the shaded parts of the circles are equal and the points on the number lines are the same distance from 0, the fractions are equivalent. Each player turns over another card. The player with the larger fraction takes all the cards from both plays. See the example on the next page.

4 The game is over when all of the cards have been taken from the deck. The player with the most cards wins.

Variation

For a greater challenge, include fraction cards from *Math Journal 2*, Activity Sheets 19–21.

Example

Jacob turns over the $\frac{3}{4}$ card and Ava turns over the $\frac{4}{6}$ card.

It is Ava's turn. She uses the circles to compare the fractions. She says that the $\frac{3}{4}$ card has a larger shaded area. Jacob takes both cards.

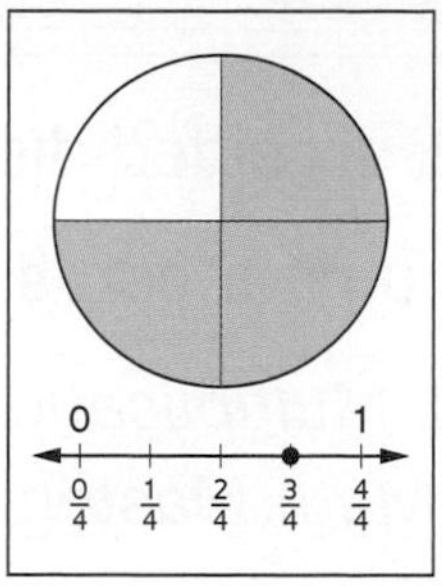

$\frac{3}{4}$

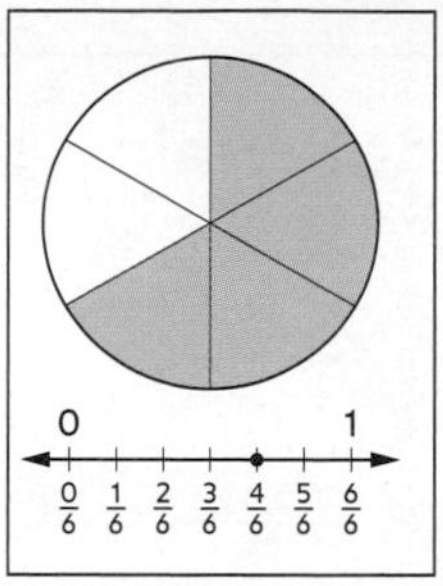

$\frac{4}{6}$

Jacob turns over the $\frac{1}{2}$ card and Ava turns over the $\frac{4}{8}$ card.

It is Jacob's turn. He uses the number lines to compare the fractions. He declares the points are the same distance from 0. Each player turns over another fraction card. The player with the larger fraction takes all of the cards.

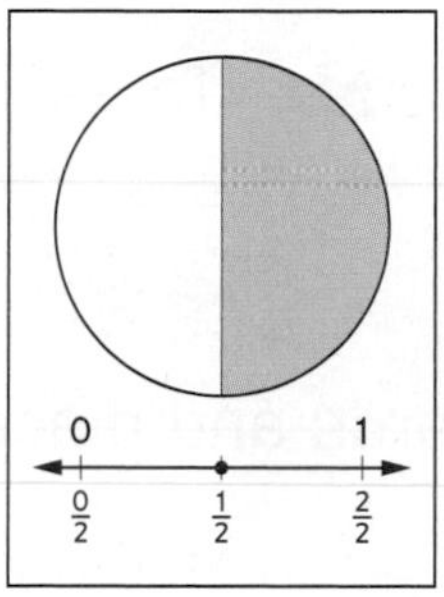

$\frac{1}{2}$

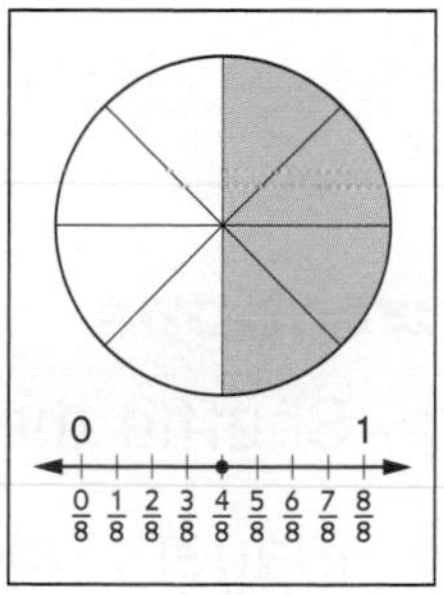

$\frac{4}{8}$

Multiplication Draw

Materials	☐ 1 six-sided die labeled with 2, 2, 5, 5, 10, and 10
	☐ number cards 1–10 (4 of each)
	☐ 1 *Multiplication Draw* Record Sheet for each player (*Math Masters*, p. G6)
Players	2 or 3
Skill	Practicing multiplication facts
Object of the Game	To have the largest score.

Directions

1 Shuffle the cards and place the deck number-side down.

2 Players take turns. When it is your turn, roll the die and draw a card from the deck to get 2 multiplication factors. Record both factors and their product on your record sheet. After 5 turns, find the sum of your 5 products. This is your score.

3 Keep playing. Add your scores for each round. The player with the largest score after 3 rounds wins the game.

Variations

• Roll a regular 6-sided die to include factors from 1 through 6.

• Draw 2 cards to get 2 multiplication factors from 1 through 10.

Example

Alex rolls a 10 and draws a 3 card. He records $10 \times 3 = 30$.

Partner 1	Round 1	Round 2	Round 3
1st draw:	_10_ × _3_ = _30_	___ × ___ = ___	___ × ___ = ___
2nd draw:	___ × ___ = ___	___ × ___ = ___	___ × ___ = ___
3rd draw:	___ × ___ = ___	___ × ___ = ___	___ × ___ = ___
4th draw:	___ × ___ = ___	___ × ___ = ___	___ × ___ = ___
5th draw:	___ × ___ = ___	___ × ___ = ___	___ × ___ = ___
Sum of products:	_____	_____	_____

Name That Number

Materials	☐ number cards 0–20 (4 of each card 0–10, and 1 of each card 11–20)
Players	2 to 4 (The game is more interesting when played by 3 or 4 players.)
Skill	Finding equivalent names for numbers
Object of the Game	To collect the most cards.

Directions

1 Shuffle the deck and place 5 cards number-side up on the table. Leave the rest of the deck number-side down. Then turn over the top card of the deck and lay it down next to the deck. The number on this card is the number to be named. This is the *target number*.

2 Players take turns. When it is your turn:

- Try to name the target number. You can name the target number by adding, subtracting, multiplying, or dividing the numbers on 2 or more of the 5 cards that are number-side up. A card may be used only once for each turn.

- If you can name the target number, take the cards you used to name it. Also take the target-number card. Then replace all the cards you took by drawing from the top of the deck. See the example on the next page.

- If you cannot name the target number, your turn is over. Turn over a new card from the top of the deck and lay it down on the target-number pile. The number on this card becomes the new target number to be named.

3 Play continues until all of the cards in the deck have been turned over. The player who has taken the most cards wins.

Example

Mae and Mike take turns.

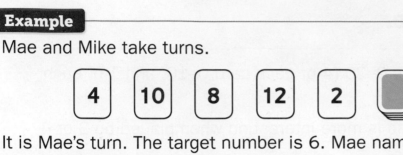

It is Mae's turn. The target number is 6. Mae names the number with 4 + 2. She could have also said 8 − 2 or 10 − 4.

Mae takes the 4, 2, and 6 cards. Then she replaces them by drawing cards from the deck.

It is now Mike's turn. The new target number is 16. Mike sees two ways to name the target number.

- He can use 3 cards and name the target number like this:

$$\boxed{7} + \boxed{8} + \boxed{1} = 16$$

- He can use 4 cards and name the target number like this:

$$\boxed{12} - \boxed{10} = 2$$
$$\downarrow$$
$$2 \times \boxed{8} = 16$$
$$\downarrow$$
$$16 \div \boxed{1} = 16$$

Mike chooses the 4-card solution because he can take more cards that way. He takes the 12, 10, 8, and 1 cards. He also takes the target-number card 16. Then he replaces all 5 cards by drawing cards from the deck.

Number-Grid Difference

Materials ☐ number cards 0–9 (4 of each)

☐ 1 Number Grid (*Math Masters*, p. TA3)

☐ 1 *Number-Grid Difference* Record Sheet for each player (*Math Masters*, p. G2)

☐ 1 counter for each player

☐ calculator (optional)

Players 2

Skill Using a number grid to find the difference between 2-digit numbers

Object of the Game To have the lower sum.

Directions

1 Shuffle the cards. Place the deck number-side down on the table.

2 Players take turns. When it is your turn:

- Each player takes 2 cards from the deck and uses the cards to make a 2-digit number. Players then place their counters on the grid to mark their numbers.

- Find the difference between your number and your partner's number.

- The difference is your score for the round. Record both numbers and your score on your record sheet.

3 Continue playing until each player has taken 5 turns and recorded 5 scores.

4 Each player finds the sum of his or her 5 scores. Players may use a calculator to add.

5 The player with the lower sum wins the game.

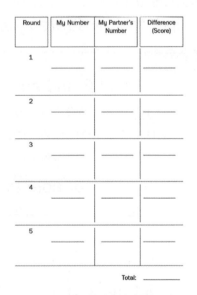

Number-Grid Difference Record Sheet

Product Pile-Up

Materials	☐ number cards 1–10 (4 of each)
Players	3
Skill	Practicing multiplication facts 1 to 10
Object of the Game	To play all of your cards.

Directions

1 Take turns being the dealer. Shuffle and deal 8 cards to each player. Place the rest of the deck number-side down.

2 The player to the left of the dealer begins. This player selects 2 cards from his or her hand, places them number-side up on the table, multiplies the numbers, and says the product aloud.

3 Play continues with each player playing 2 cards with a product *greater than* the product of the last 2 cards played. If a player states an incorrect product, other players may suggest a helper fact or strategy to help find the correct product.

> **Example**
> Joe plays 3 and 6 and says, "3 times 6 equals 18." Then Rachel looks at her hand to find two cards with a product greater than 18. She plays 5 and 4 and says, "5 times 4 equals 20."

4 If a player is not able to play 2 cards with a greater product, the player draws 2 cards from the deck.

- If the player is now able to make a greater product, those cards are played and the game continues. If the player still cannot make a greater product, the player keeps the cards and says "Pass." The game continues to the next person.

- If all players must pass, the player who laid down the last 2 cards starts a new round beginning with Step 2 above.

5 The winner is the first player to run out of cards, or the player with the fewest cards when there are no more cards to draw.

Roll to 1,000

Materials	☐ 1 *Roll to 1,000* Record Sheet (*Math Masters*, p. G7)
	☐ 2 six-sided dice
Players	2 to 4
Skill	Adding multiples of 10
Object of the Game	To score at least 1,000.

Directions

Each dice roll represents a number of tens. For example, if you roll a 3 and a 4 for a total of 7, you have 7 tens, or 70.

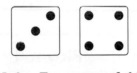

Make 7 groups of 10, or 70.

1 Players take turns. When it is your turn:

- Roll the dice as many times as you want. Each roll tells you how many tens you have.

- Mentally add the numbers you get for all of your dice rolls. Enter this as your score for the turn.

- If you roll a 1, your turn is over. Enter 0 as your score for this turn.

2 Continue to add to your score each turn. If you roll a 1 at any time, your score for that turn is 0. The score you enter is the total from your previous turn. See the example on the next page.

3 The first player to score 1,000 or more wins the game.

Turn	Player 1	Player 2	Player 3	Player 4
1				
2				
3				
4				
5				
6				
7				
8				
9				
10				
11				
12				
13				
14				
15				

***Roll to 1,000* Record Sheet**

Example

- On his first turn, Ray rolls a 4 and a 5, which is 9 tens, or 90. He decides to stop rolling, so his score for Turn 1 is 90.

- On his second turn, Ray rolls a 4 and a 2, which is 6 tens, or 60. He decides to keep rolling and gets a 3 and a 1. Because he rolled a 1, Ray's turn is over and his score for Turn 2 is 0. He keeps his score from Turn 1, 90, and records it again as his score for Turn 2.

- On his third turn, Ray rolls a 6 and a 6, which is 12 tens, or 120. He rolls again and gets a 3 and a 2. He adds 50 to 120 to get 170. He decides to stop rolling, so he adds 170 to his Turn 2 score of 90. His score for Turn 3 is 90 + 170 = 260.

Turn	Player 1 ____Ray____	Player 2 _____	Player 3 _____	Player 4 _____
1	90			
2	90			
3	260			

Variations

Roll to 500: For a shorter game, play to 500.

Roll to 100: For an easier game, use the sums of the actual dice rolls. The object of the game is to be the first to reach 100.

Roll and Multiply to 1,000: Roll 2 dice. One roll represents the number of tens. Multiply the number of tens by the other roll. For example, if you roll a 3 and a 4, multiply 30 by 4 to get 120.

Back to Zero: Play any of the versions above. A player who reaches or exceeds the goal continues to take turns, but subtracts the numbers rolled each time instead of adding them. The first player to get back to zero or less wins.

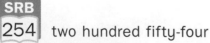

Salute!

Materials	☐ number cards 1–6 and 10 (4 of each)
Players	3
Skill	Practicing multiplication and division facts
Object of the Game	To solve for a missing factor.

Directions

1. One person begins as the "Dealer." The Dealer gives one card to each of the other two "Players."

2. Without looking at their cards, the Players hold them on their foreheads with the numbers facing out.

3. The Dealer looks at both cards and says the product of the two numbers.

4. Each Player looks at the other Player's card. They use the number they see and the product said by the Dealer to figure out the number on their card (the missing factor). They say that number out loud.

5. Once both Players have said their numbers, they can look at their own cards to check their answers.

6. Rotate roles clockwise and play again.

7. Play continues until everyone has been the Dealer five times.

The product is 15.

The players use the number they can see and the product to figure out the number on their own card.

Variations

- Play with number cards 1–10 (4 of each).

- To practice addition and subtraction facts, the Dealer looks at both cards and says the sum of the two numbers. Each Player looks at the other Player's card. They use the number they see and the sum said by the Dealer to figure out the number on their card (the missing addend). They say that number out loud.

Shuffle to 100

Materials   ☐ number cards 1–9 (4 of each)

☐ 1 *Shuffle to 100* Record Sheet for each player
(*Math Masters*, p. G10)

Players   2 or 3

Skill   Estimating sums and making combinations close to 100

Object of the Game   To have the lowest score.

Directions

1 Take turns being the dealer. The dealer shuffles the cards and deals 5 number cards to each player number-side up.

2 Players think about two 2-digit numbers they can make with their cards. For example, the cards 3 and 4 could make either 34 or 43.

3 Players estimate to decide which of their possible 2-digit numbers will add up to a sum closest to 100. After choosing four cards to use, each player discards the extra card.

4 Players add their two 2-digit numbers to find an exact answer.

5 Players record their 2-digit numbers and sums on their *Shuffle to 100* Record Sheet.

6 Each player finds his or her score by finding the difference between their sum and 100.

7 Players discard the cards they used. When all of the cards have been used, the dealer shuffles the discarded cards to make a new deck to finish the game.

Score

Round 1: _____ + _____ = _____   _____

Round 2: _____ + _____ = _____   _____

Round 3: _____ + _____ = _____   _____

Round 4: _____ + _____ = _____   _____

Total Score: _____

Shuffle to 100 Record Sheet

8 The next dealer deals 5 new cards to each player. Players make new combinations that will add to a sum closest to 100.

9 Players add their scores after 4 rounds. The player with the LOWEST total score wins.

Example

Steve is dealt an 8, 9, 5, 7, and 2.
He makes 75 and 28 and discards
the 9. 75 + 28 = 103, so his score
is 3, because the difference between
100 and 103 is 3.

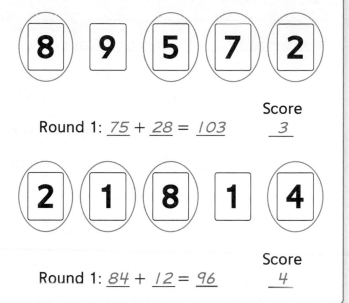

Round 1: <u>75</u> + <u>28</u> = <u>103</u>

Score
<u>3</u>

Bobby is dealt a 2, 1, 8, 1, and 4.
He makes 84 and 12 and discards
the 1. 84 + 12 = 96, so his score is
4, because the difference between
100 and 96 is 4.

Round 1: <u>84</u> + <u>12</u> = <u>96</u>

Score
<u>4</u>

Variation

Shuffle to 1,000: Play the game as directed, except the
dealer deals 7 cards to each player. Players make two
3-digit numbers and discard the extra card. Players record
their numbers and sums on the *Shuffle to 1,000* Record Sheet
(*Math Masters*, page G11). Then players find and record the
difference between their sums and 1,000. The player with the
LOWEST total score after 4 rounds wins.

Spin and Round

Materials
☐ 1 *Spin and Round* Spinner (*Math Masters*, p. G4)

☐ 1 *Spin and Round* Record Sheet for each player (*Math Masters*, p. G5)

☐ number cards 1–9 (4 of each)

☐ 1 pencil and 1 large paperclip for the spinner

Players 2 or 3

Skill Rounding numbers to the nearest 10 or 100

Object of Game To have the largest total sum.

Directions

1 Shuffle the cards and place the deck number-side down.

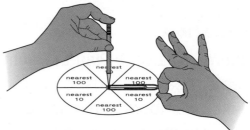

2 Players take turns. When it is your turn:

• Spin the spinner and record on your *Spin and Round* Record Sheet whether you must round to the nearest 10 or the nearest 100.

• Take 3 cards from the deck and make a 3-digit number.

• Round the 3-digit number to the nearest 10 or the nearest 100, depending on your spin.

• Record both the original number and the rounded number on your record sheet. See the example on the next page.

3 After 5 turns, each player finds the total sum of their rounded numbers.

4 The player with the largest total sum wins the game.

Variation

Take 4 cards and make 4-digit numbers.

Example

Turn 1: Tiffany spins and lands on "nearest 10." She draws a 3, 5, and 2.
She makes the number 532 and rounds it to 530.

Turn 2: Tiffany spins and lands on "nearest 100." She draws a 1, 2, and 4.
She makes the number 412 and rounds it to 400.

Turn	Nearest 10? Nearest 100?	Original Number	Rounded Number
Example	10	536	540
1	10	532	530
2	100	412	400

I know that 532 is closer to 530 than 540, so the nearest 10 is 530.

Top-It Games

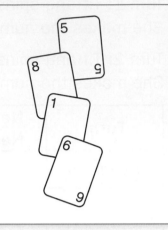

Materials ☐ number cards 0–10 (4 of each)

Players 2 to 4

Skill Practicing basic facts

Object of the Game To collect the most cards.

Multiplication Top-It

Directions

1. Shuffle the cards. Place the deck number-side down on the table.

2. Each player turns over 2 cards and calls out the product of the numbers.

3. The player with the largest product wins the round and takes all the cards.

4. In case of a tie for the largest product, each tied player turns over 2 more cards and calls out the product of the numbers. The player with the largest product then takes all the cards from both plays.

5. The game ends when there are not enough cards left for each player to have another turn.

6. The player with the most cards wins.

Example

Round 1:

• Ann turns over a 2 and a 6.
  She multiplies 2 × 6 and calls out 12.

• Beth turns over a 6 and a 0.
  She multiplies 6 × 0 and calls out 0.

• Joe turns over a 10 and a 4.
  He multiplies 10 × 4 and calls out 40.

Joe has the largest product. He takes all 6 cards.

2	6
6	0
10	4

Example

Round 2:

- Ann turns over a 3 and an 8. She calls out 24.

- Beth turns over a 4 and a 6. She calls out 24.

- Joe turns over a 9 and a 2. He calls out 18.

Ann and Beth are tied with 24, so they each turn over 2 more cards.

- Ann turns over a 3 and a 7. She calls out 21.

- Beth turns over an 8 and a 4. She calls out 32.

Beth wins Round 2 and takes all 10 cards.

Multiplication Top-It (with Extended Facts)

Directions

Multiplication Top-It (with Extended Facts) is played like *Multiplication Top-It*, except players make the second card a multiple of 10. For example, a player turns over a 2 and then a 6. The player uses 60 instead of 6 and multiplies $2 \times 60 = 120$. All other directions stay the same.

Addition Top-It

Directions

Addition Top-It is played like *Multiplication Top-It*, except players call out the sum of the 2 numbers. The player with the largest sum wins the round and takes all the cards.

Subtraction Top-It

Directions

Subtraction Top-It is played like *Multiplication Top-It*, except players subtract the smaller number from the larger number and call out the difference. The player with the largest difference wins the round and takes all the cards.

What's My Polygon Rule?

Materials	☐ Shape Cards 1 and 2 (*Math Masters*, pp. G13–G14)
	☐ 1 sheet of paper
Players	2 or more
Skill	Understanding attributes of polygons
Object of the Game	To guess the rule.

Directions

1 Spread out the shape cards faceup on a table. Draw a large circle on a sheet of paper.

2 Players take turns. When it is your turn as the Rule Maker:

- Think of a rule. Do not say your rule out loud. Choose two shapes that fit your rule. Put those shapes inside the circle. Choose at least one shape that does not fit your rule and put that shape outside the circle.

- Ask the other players to choose one other shape that they think fits your rule. If it fits your rule, add it to the circle. If it does not fit, place it outside the circle.

- Ask the other players to guess your rule. Once someone correctly guesses the rule, it is another player's turn to be the Rule Maker.

Example

Coby places his shapes in the following way:

Jayla thinks Shape B fits the rule, and Coby adds it to the circle. Jayla says, "I think the rule is polygons with 3 sides."

Coby confirms that Jayla correctly guessed the rule.

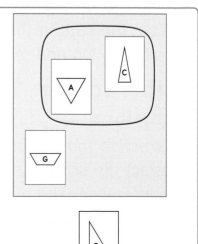

POLO SUR 41

TORONTO

MADRID 12.

SIDNEY 9.485 KM

ROMA 13.158 KM

MIAM

16.311 KM

.94 KM

LONDRES 13.387 KM

AMSTERDAM 13.688

RIO JANEIRO 4.101 KM

ATENAS 13.512 KM

MOSCU 15.573 KM

SANTIAGO-CHILE 219

KORE

17.797 kms.

Introduction

Sharing information is an important part of learning. Scientists, historians, mathematicians, and others collect data by measuring, counting, and making observations of objects and people. After organizing and analyzing the data, they share it with others, so that everyone can learn from their observations and findings.

Sometimes there is so much information that it can be hard to find the data you need. This section is a collection of interesting data that you can use to explore and practice what you are learning in math.

Some of the information in this section may be easy to find on your own because it comes from one place. For example, you can find out the normal spring temperatures for many cities in the United States by going to the National Oceanographic and Atmospheric Administration's (NOAA) website. It gives many kinds of data about the weather in different parts of the country.

Some of the information comes from many different places. It could take you a long time to find the information on your own. For example, to learn the masses of balls for different sports, you would have to search each sport's official website or find books about them. Instead, the information has been collected and organized for you here.

The places where you find data are called sources. Pages 289–290 list sources used to find the data in this section.

(tr)McGraw-Hill Education; (cr)Purestock/SuperStock; (br)René Mansi/E+/Getty Images; (bl)McGraw-Hill Companies Inc./Ken Karp, photographer

Normal Spring High and Low Temperatures (in °F)

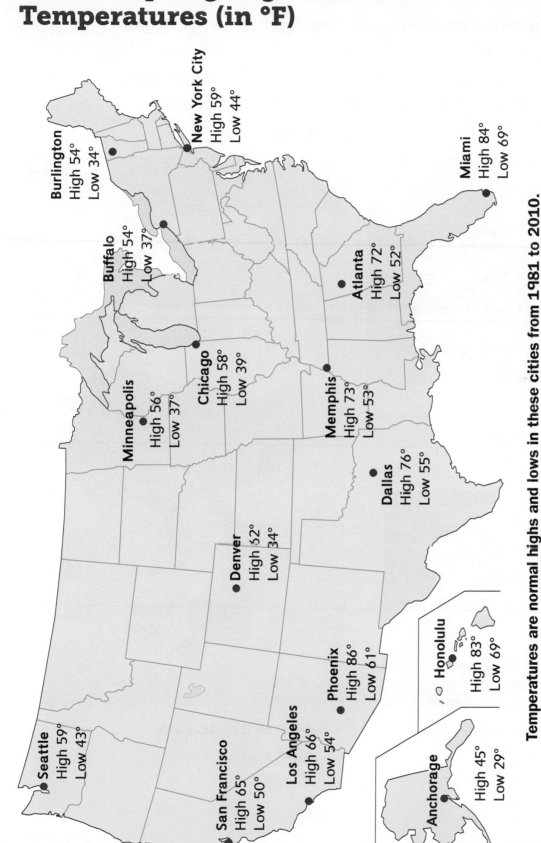

Burlington
High 54°
Low 34°

New York City
High 59°
Low 44°

Miami
High 84°
Low 69°

Buffalo
High 54°
Low 37°

Atlanta
High 72°
Low 52°

Chicago
High 58°
Low 39°

Minneapolis
High 56°
Low 37°

Memphis
High 73°
Low 53°

Dallas
High 76°
Low 55°

Denver
High 52°
Low 34°

Phoenix
High 86°
Low 61°

Honolulu
High 83°
Low 69°

Los Angeles
High 66°
Low 54°

Seattle
High 59°
Low 43°

San Francisco
High 65°
Low 50°

Anchorage
High 45°
Low 29°

Temperatures are normal highs and lows in these cities from 1981 to 2010.
70°F is normal room temperature.

Animal Clutches

All of the animals shown on these pages lay eggs. A nest of eggs is called a *clutch*.

Most birds, reptiles, and amphibians lay eggs once or twice a year. Insects may lay eggs daily during a certain season of the year.

Sea turtle

- **up to 1.5 meters long**
- **between 70 and 190 eggs**

Giant toad

- **up to 25 centimeters long**
- **around 35,000 eggs**

Ostrich

- **more than 2 meters tall**
- **about 15 eggs**

Python

- up to 9 meters long
- up to 100 eggs

Agama lizard

- up to 25 centimeters long
- up to 20 eggs

Queen termite

- about 1 centimeter long
- around 8,000 eggs per day for years

Mississippi alligator

- more than 2 meters long
- as many as 88 eggs

Drinks Vending Machine Poster

Example

How much money do you need to buy both a strawberry yogurt drink and an apple juice?

A strawberry yogurt drink costs 70¢ and an apple juice costs 55¢.

70¢ + 55¢ = ?

70¢ + 55¢ = 125¢. You need 125¢ to buy both drinks.

Another way to think about 125¢ is as 100¢ + 25¢. 100¢ is the same as 1 dollar, or $1.00. So to buy both drinks, you need $1.00 and 25¢ more, or $1.25.

Example

If you wanted to buy orange juice using exact change, which coins could you use?

Orange juice costs 65¢. One quarter equals 25¢, so two quarters equal 50¢.

65¢ = 50¢ + ?

65¢ = 50¢ + 15¢ You can make 15¢ using a dime (10¢) and a nickel (5¢).

So to buy orange juice using exact change, you could use two quarters (25¢ + 25¢), one dime (10¢), and one nickel (5¢).

Snacks Vending Machine Poster

Dollar Store Poster

Toys

- **Mini stock cars**
  10 per box   **$4** per box

- **Marbles**
  45 per bag   **$2** per bag

- **Blocks**
  395 pieces   **$20** per set

Fashion

- **Shoelaces**
  5 pairs per package   **$3** per pkg.

- **Ponytail rings**
  12 per package   **$2** per pkg.

- **Hair clips**
  6 per bag   **$1** per bag

School Supplies

- **Notebook paper**
  200 sheets   **$1** per pkg.
  per package

- **Value-pack pens**
  10 in a package   **$1** per pkg.

- **Chocolate-scented pens**
  6 in a pack   **$2** per pack

- **Fashion pens**
  4 in a pack   **$2** per pack

- **File cards**
  package of 100 for **$2**

- **Brilliant-color markers**
  package of 5 for **$2**

- **Scented markers**
  package of 8 for **$3**

- **Pencils**
  8-pack **$2**

Party Supplies

- **Glitter stickers**
  7 per pack   **$1** per pack

- **9-inch balloons**
  25 per bag   **$2** per bag

- **Party hats**   6 for **$2**
- **Party horns**   8 for **$3**
- **Giant 14-inch balloons**
  package of 5 for **$3**

Masses of Sports Balls

Mr. Isaacs, the Lincoln School physical education teacher, had students in the school measure the mass of several balls from the gym. The table below lists the masses of the sports balls. Each mass is given in grams.

Masses of Sports Balls	
Ball	Mass (in grams)
Table tennis	$2\frac{1}{2}$ g
Squash	25 g
Golf	43 g
Tennis	57 g
Baseball	142 g
Cricket	156 g
Softball	184 g
Volleyball	270 g
Soccer	425 g
Water polo	425 g
Croquet	454 g
Basketball	625 g
Bowling	7,260 g

Note I kilogram equals I,000 grams.

A wooden baseball bat has a mass of about I kilogram.

Check Your Understanding

1. Which sports balls have a mass that is 1 kilogram or more?

2. Which ball's mass is closest to 200 grams?

3. How much more mass does a basketball have than a volleyball?

Check your answers in the Answer Key.

U.S. Road Mileage Map

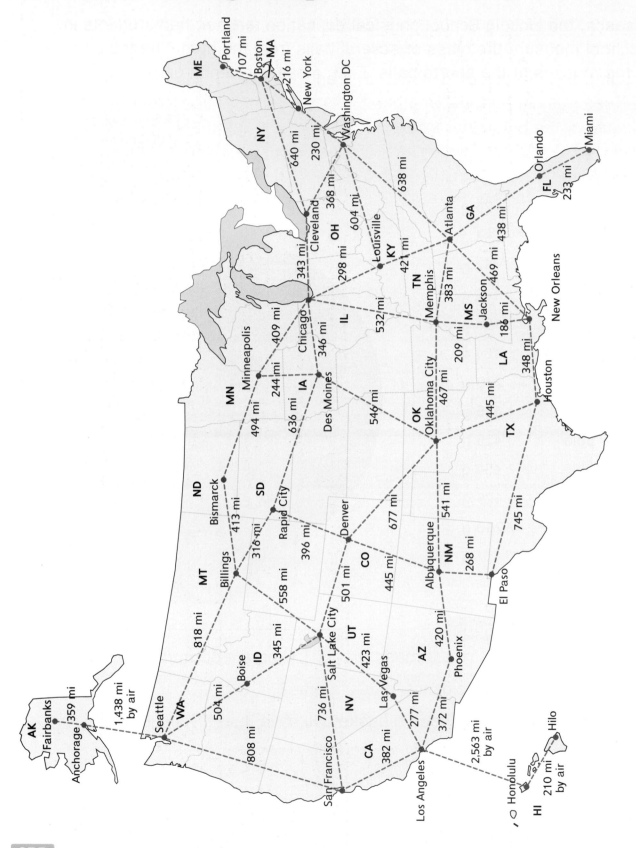

Except for the distances shown for Hawaii and Alaska, all numbers are highway distances in miles.

Information about North American Birds

The area of North America is about 9,400,000 square miles. North America is $2\frac{1}{2}$ times larger than the United States.

Although the birds on journal pages 277 and 278 are positioned on or near areas they inhabit, their natural habitats may be much larger than the locations shown. Some birds may live in and move about a very large area, or *range*. Birds that are *endangered* or have small populations may have a much smaller range. Birds often have different ranges in summer and winter if they *migrate* to breed. Here is more information about these birds.

American white pelican

Range: Canada and the interior United States in summer (breeding season); and the southern United States, Mexico, and Central America in winter
Habitats: Islands in interior freshwater lakes, bays, and estuaries, but not open seashore

Atlantic puffin

Range: Coastal land along the North Atlantic Ocean, from Greenland and Iceland to Norway and Newfoundland; Canada in summer (breeding season); and in the open ocean as far south as the Atlantic Ocean along the northern United States and Spain in winter
Habitats: Coastal cliff tops for breeding, and open ocean

Bald eagle

Range: Most of Canada and Alaska in summer (breeding season), and throughout the United States and Mexico in winter; some year-round populations along the coast of Alaska, the western coast of Canada, and the northwestern portion of the United States
Habitats: Large bodies of water, such as lakes, reservoirs, rivers, marshes, and coasts

Black-capped chickadee

Range: Year-round in most of Canada, Alaska, and throughout the northern United States, as far south as Colorado and Tennessee

Habitats: Trees or woody bushes, forests, wooded areas of areas populated by humans; often found in alder and birch trees

Blue jay

Range: Eastern and central United States and southern Canada, from Maine to Colorado and Wyoming; some migratory populations breed as far north and west as northern Alberta, Canada, but in winter return south to the United States

Habitats: Forest edges, often near oak trees to feed on acorns

Blue-crowned motmot

Range: Year-round from Mexico through Central America and most of northern South America, including Venezuela, Colombia, Guyana, and northern Brazil and Bolivia

Habitats: Forests and woodlands

California condor

Range: Northern Arizona, southern Utah, central and southern California; considered critically endangered (very few birds still exist in the wild)

Habitats: Cliffs and forests in mountains up to 6,000 feet; also along beaches and meadows to look for food

Great blue heron

Range: Throughout Canada and the United States in summer (breeding season); United States, Mexico, and into Central and South America (Honduras and Panama, and northern Colombia, Venezuela, and Ecuador) in winter

Habitats: Bodies of saltwater and freshwater, such as coastlines, marshes, rivers, lakes, or even backyard ponds

Great horned owl

Range: Year-round throughout North America, Central America, and most parts of South America, except for the Amazon basin (in northern Brazil)

Habitats: Wooded areas, often close to fields or other open areas

House sparrow

Range: Worldwide, including all of Europe; most of eastern and central Asia; northern and southern Africa; most of North and Central America; parts of South America, including eastern Brazil, Paraguay, Uruguay, and Argentina; and eastern Australia

Habitats: Near where humans live, in trees, grasses, and bushes; often nest and breed in buildings, such as factories or warehouses

Killdeer

Range: Most of Canada, the continental United States, and northern Mexico in summer (breeding season); the United States, Mexico, Central America, and into northern Colombia, Venezuela, and Ecuador in winter

Habitats: Areas of low vegetation, such as grass, sandbars, or mudflats

Ladder-back woodpecker

Range: Year-round in the southern United States (Texas, New Mexico, Arizona, and parts of California), throughout Mexico, and the eastern coast of Nicaragua

Habitats: Dry areas of brush and trees

Northern cardinal

Range: Year-round in the eastern and central parts of the United States, from Maine to Texas and eastern South Dakota, and sometimes as far west as Arizona; and the northern and eastern parts of Mexico, into the Yucatan Peninsula

Habitats: Shrubs and forests, often around where humans live

Ruby-throated hummingbird

Range: Eastern half of the United States and southern Canada in summer (breeding season); southern Mexico, Central America as far south as Costa Rica, and the coasts of southern Florida in winter

Habitats: Forest edges and gardens

Snowy owl

Range: Along the coasts in extreme northern Canada and Alaska, Europe, and Asia in summer (breeding season); throughout most of northern Europe, Asia, and North America, traveling as far south as Maine and Vermont in winter

Habitats: Arctic tundra for breeding; lake and ocean shores

Yellow-headed parrot

Range: Small parts of the eastern and western coasts of Mexico, Belize, and the eastern tip of Guatemala; considered endangered (not many birds still exist in the wild)

Habitats: Forested areas along water, evergreen forests, and mangrove forests

Heights of 8-Year-Old Children

The table below shows the heights of the children in a third-grade class, measured to the nearest centimeter and to the nearest $\frac{1}{4}$ inch. All of the children who were measured were 8 years old.

Heights of 8-Year-Old Children					
Boys			Girls		
Boy	Height (in cm)	Height (in in.)	Girl	Height (in cm)	Height (in in.)
1	136 cm	$53\frac{1}{2}$ in.	1	123 cm	$48\frac{1}{2}$ in.
2	129 cm	$50\frac{3}{4}$ in.	2	141 cm	$55\frac{1}{2}$ in.
3	110 cm	$43\frac{1}{4}$ in.	3	115 cm	$45\frac{1}{4}$ in.
4	122 cm	48 in.	4	126 cm	$49\frac{1}{2}$ in.
5	126 cm	$49\frac{1}{2}$ in.	5	122 cm	48 in.
6	148 cm	$58\frac{1}{4}$ in.	6	144 cm	$56\frac{3}{4}$ in.
7	127 cm	50 in.	7	127 cm	50 in.
8	126 cm	$49\frac{1}{2}$ in.	8	133 cm	$52\frac{1}{4}$ in.
9	124 cm	$48\frac{3}{4}$ in.	9	120 cm	$47\frac{1}{4}$ in.
10	142 cm	56 in.	10	125 cm	$49\frac{1}{4}$ in.
11	118 cm	$46\frac{1}{2}$ in.	11	126 cm	$49\frac{1}{2}$ in.
12	130 cm	$51\frac{1}{4}$ in.	12	107 cm	$42\frac{1}{2}$ in.

What are some things you can say about the heights of 8-year-old boys and girls?

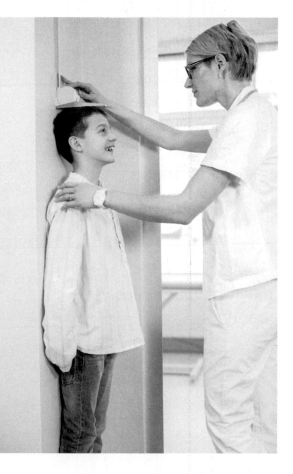

Lengths of Day for World Cities

The length of day can be very different depending on where you are on Earth. The hours of sunlight depend on the time of year and how close a location is to the equator because:

- Earth makes one full revolution around the sun each year; and

- Earth does not sit straight up and down, but sits at a slight angle, or tilt.

From June to September, when it is summer in the Northern Hemisphere, it is winter in the Southern Hemisphere. During this time, places in the Northern Hemisphere, such as Iceland, get more hours of sunlight during their summer. During the same time, places in the Southern Hemisphere, such as Tierra del Fuego, get fewer hours of sunlight during their winter. As Earth revolves around the sun, the Northern Hemisphere gets fewer hours of sunlight during its winter (December to March). The Southern Hemisphere gets more hours of sunlight during its summer (December to March).

In places farther from the equator, the number of hours of sunlight each day varies greatly from summer to winter. For example, in Iceland the number of hours of sunlight in the summer is much greater than the number of hours of sunlight in the winter. The length of day in places near the equator does not change much as Earth revolves around the sun.

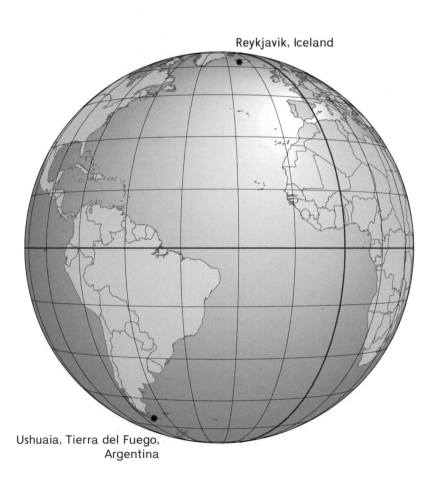

Reykjavik, Iceland

Ushuaia, Tierra del Fuego, Argentina

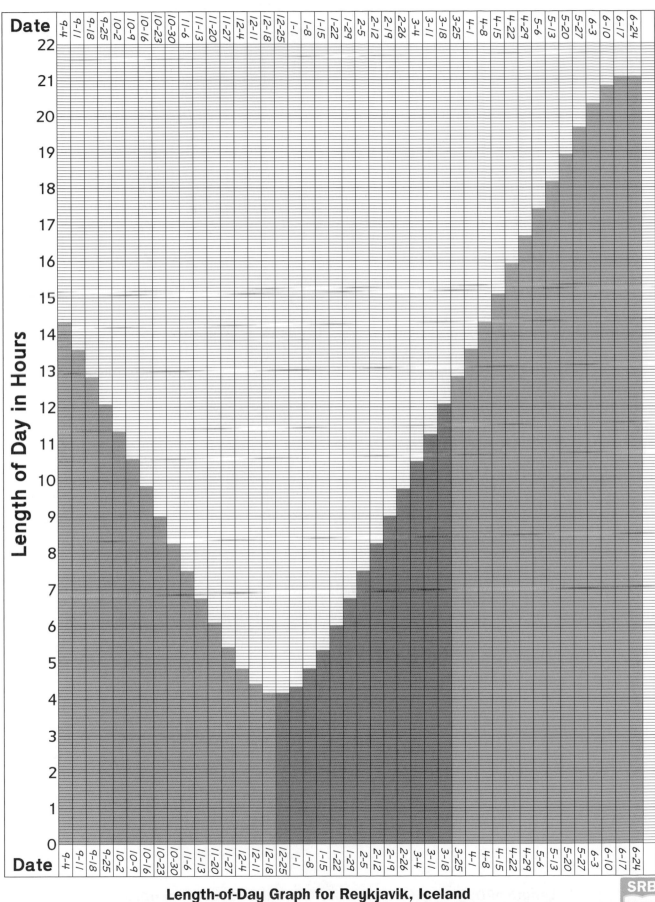

Length-of-Day Graph for Reykjavik, Iceland

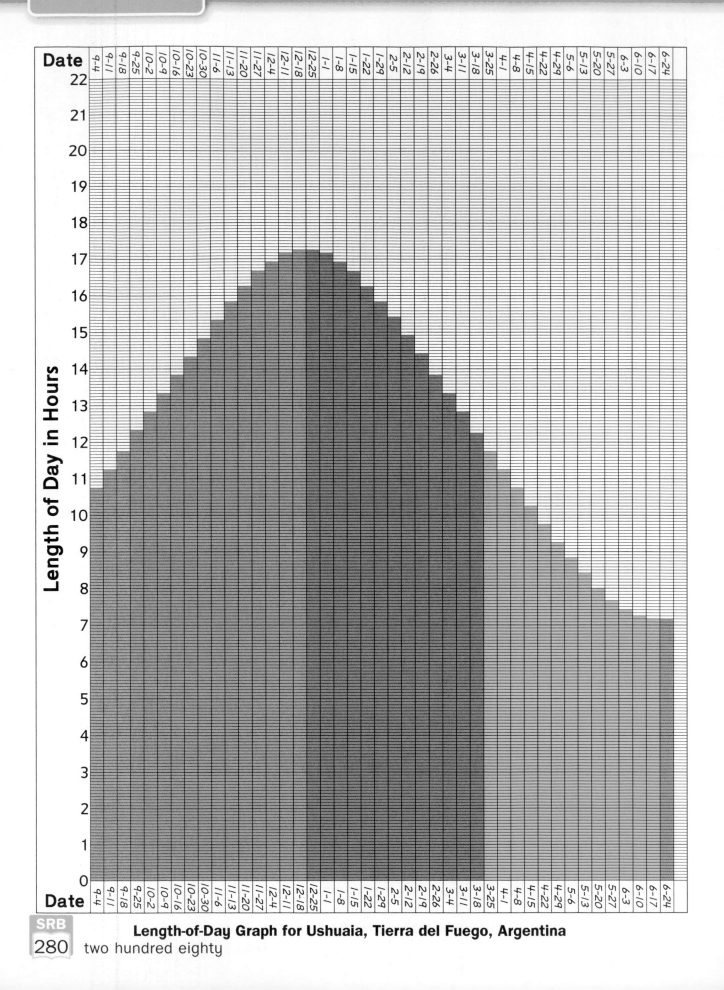

Length-of-Day Graph for Ushuaia, Tierra del Fuego, Argentina

Sunrise and Sunset Data for June 21, 2016

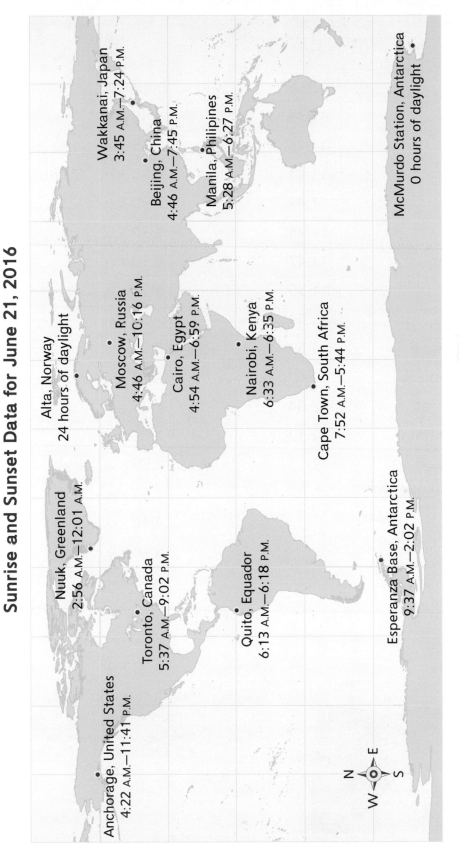

Sunrise and Sunset Data for June 21, 2016

Anchorage, United States
4:22 A.M.—11:41 P.M.

Nuuk, Greenland
2:56 A.M.—12:01 A.M.

Toronto, Canada
5:37 A.M.—9:02 P.M.

Quito, Equador
6:13 A.M.—6:18 P.M.

Esperanza Base, Antarctica
9:37 A.M.—2:02 P.M.

Alta, Norway
24 hours of daylight

Moscow, Russia
4:46 A.M.—10:16 P.M.

Cairo, Egypt
4:54 A.M.—6:59 P.M.

Nairobi, Kenya
6:33 A.M.—6:35 P.M.

Cape Town, South Africa
7:52 A.M.—5:44 P.M.

Wakkanai, Japan
3:45 A.M.—7:24 P.M.

Beijing, China
4:46 A.M.—7:45 P.M.

Manila, Philipines
5:28 A.M.—6:27 P.M.

McMurdo Station, Antarctica
0 hours of daylight

N E S W

Train Schedule and Airline Schedule

Train Schedule for a Chicago Rail Line	
Station	Time
Kensington	11:34 A.M.
Pullman	11:36
Chesterfield	11:41
Avalon Park	11:44
Chatham	11:45
Grand Crossing	11:47
63rd Street	11:50
57th Street	11:53
McCormick Place	11:59
Museum Campus	12:02 P.M.
Van Buren Street	12:04
Millennium Station	12:12

Airline Schedule from Chicago to New York City	
Departure	Arrival
5:45 A.M.	11:51 A.M.*
5:55 A.M.	9:00 A.M.
6:40 A.M.	9:45 A.M.
7:30 A.M.	10:35 A.M.
7:30 A.M.	1:41 P.M.*
8:30 A.M.	11:40 A.M.
9:25 A.M.	12:29 P.M.
10:30 A.M.	1:40 P.M.
12:00 P.M.	3:15 P.M.
12:10 P.M.	5:59 P.M.*
1:00 P.M.	4:15 P.M.
2:05 P.M.	5:15 P.M.
2:55 P.M.	6:10 P.M.
3:45 P.M.	7:00 P.M.
4:40 P.M.	8:05 P.M.
5:40 P.M.	9:05 P.M.
6:35 P.M.	9:50 P.M.
7:40 P.M.	10:50 P.M.
8:00 P.M.	11:10 P.M.
8:30 P.M.	11:40 P.M.

*Flight makes other stops.

Times shown are local times.

New York City is in the Eastern Time Zone. It is 1 hour ahead of Chicago, which is in the Central Time Zone.

Number of Words Children Know

When babies are about 1 year old, they begin to use and understand a few words.

By the time a child is 6 years old, he or she understands and uses around 2,000 to 3,000 words.

The table below shows the number of words that one child understood and used as she got older. A **linguist,** a person who studies language, collected these results as she observed and recorded her own child learning to speak.

Number of Words a Child Understood and Used	
Age (in years)	Number of Words
1	3
$1\frac{1}{2}$	87
2	307
$2\frac{1}{2}$	570
3	956
$3\frac{1}{2}$	1,422
4	1,740
$4\frac{1}{2}$	2,072
5	2,369
6	2,824

Did You Know?

Experts at the word game Scrabble typically know about 70,000–80,000 words. That's about 2 to $2\frac{1}{2}$ times as many words as the average adult knows.

Check Your Understanding

How many new words did this child begin to use between the child's second and third birthdays?

Check your answer in the Answer Key.

Letter Frequencies

There are 26 letters in the English alphabet. Some letters (such as E and T) are used quite often. Other letters (such as Q and Z) are not used very much.

The table below shows how often each letter might be used if you looked at 1,000 letters in a typical set of English words. If you wrote enough English words to write 1,000 letters, you could expect to use about this number of each letter.

Letter Frequencies	
82 As	70 Ns
14 Bs	80 Os
28 Cs	20 Ps
38 Ds	1 Q
130 Es	68 Rs
30 Fs	60 Ss
20 Gs	105 Ts
53 Hs	25 Us
65 Is	9 Vs
1 J	15 Ws
4 Ks	2 Xs
34 Ls	20 Ys
25 Ms	1 Z

English Vowel Frequencies

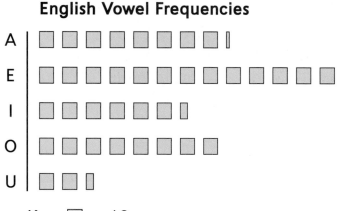

Key: ☐ = 10 occurrences

The picture graph above shows the number of times each English vowel will occur in a typical set of 1,000 letters.

Check Your Understanding

1. Which 5 letters are used the most?

2. Which 5 letters are used the least?

3. The letters A, E, I, O, and U are called *vowels*. How many vowels are in your full name (first name and last name)? Do you have more Es or Us in your name?

Check your answers in the Answer Key.

How Much Would You Weigh on the Moon?

Mass is a measure of the amount of matter in an object. **Weight** is a measure of the force of gravity given an object's mass. All objects are attracted to each other by a force called *gravity*. Gravity pulls your body toward the center of Earth.

When you weigh yourself on a scale on Earth, the scale measures the pull of Earth's gravity on the mass of your body. That is your *weight*. If you traveled to another planet or to the moon, your *mass* would not change because the amount of matter in your body does not change.

However, the pull of the moon's gravity is much less than the pull of Earth's gravity. You would weigh less on the moon than on Earth. Jupiter's gravity is stronger than Earth's gravity. You would weigh more on Jupiter than on Earth.

The table below shows what a person's weight would be if he or she traveled to the planets, the sun, and the moon. Remember, as the person travels from planet to planet, his or her mass stays the same.

Location	Person's Weight
Earth	100 pounds
Moon	17 pounds
Sun	2,800 pounds
Mercury	38 pounds
Venus	90 pounds
Mars	38 pounds
Jupiter	236 pounds
Saturn	106 pounds
Uranus	89 pounds
Neptune	112 pounds

Physical Fitness Standards

The table on the next page shows data for three fitness tests.

Curl-Ups

A partner holds your feet. You cross your arms and place your hands on opposite shoulders. You raise your body and curl up to touch your elbows to your thighs. Then you lower your back to the floor. This counts as one curl-up. Do as many curl-ups as you can in one minute.

One-Mile Run/Walk

Cover a 1-mile distance in as short a time as you can. You may not be able to run the entire distance. Walk when you are not able to run.

Arm Hang

Hold the bar with your palms facing away from your body. Your chin should clear the bar. (See the picture.) Hold this position as long as you can.

Did You Know?

In order to be physically fit, children between the ages of 7 and 11 need at least 10 hours of sleep each night.

Physical Fitness Test Scores (median scores for each age)

	Age	Curl-Ups (in 1 minute)	One-Mile Run/Walk (minutes:seconds)	Arm Hang (seconds)
Boys	6	22	12:36	6
	7	28	11:40	8
	8	31	11:05	10
	9	32	10:30	10
	10	35	9:48	12
	11	37	9:20	11
	12	40	8:40	12
Girls	6	23	13:12	5
	7	25	12:56	6
	8	29	12:30	8
	9	30	11:52	8
	10	30	11:22	8
	11	32	11:17	7
	12	35	11:05	7

The data in the table are **median** scores for each age group. When you place all of the results from children in that age group in order from least to greatest, or shortest to longest, the median is the *middle* value. That means about half of all of the children had results that were *less* than the number in the table, and about half of the children's results were *more* than the number in the table.

Example

The table shows 31 curl-ups for 8-year-old boys. 31 curl-ups is the *median* score for 8-year-old boys. About half of all 8-year-old boys will do *more than* 31 curl-ups, and about half of them will do *fewer than* 31 curl-ups.

Example

The table shows a *median* time of 12 minutes and 30 seconds for 8-year-old girls in the mile run. About half of all 8-year-old girls will take *more than* 12 minutes and 30 seconds to run a mile, and about half of all 8-year-old girls will take *less than* 12 minutes and 30 seconds.

Tables of Measures

Metric System

Units of Length	Units of Area
1 kilometer (km) = 1,000 meters (m) 1 meter (m) = 10 decimeters (dm) = 100 centimeters (cm) = 1,000 millimeters (mm) 1 decimeter (dm) = 10 centimeters (cm) 1 centimeter (cm) = 10 millimeters (mm)	1 square meter = 10,000 square centimeters 1 sq m = 10,000 sq cm 1 square centimeter = 100 square millimeters 1 sq cm = 100 sq mm

Units of Mass	Units of Liquid Volume
1 metric ton (t) = 1,000 kilograms (kg) 1 kilogram (kg) = 1,000 grams (g) 1 gram (g) = 1,000 milligrams (mg)	1 kiloliter (kL) = 1,000 liters (L) 1 liter (L) = 1,000 milliliters (mL)

U.S. Customary System

Units of Length	Units of Area
1 mile (mi) = 1,760 yards (yd) = 5,280 feet (ft) 1 yard (yd) = 3 feet (ft) = 36 inches (in.) 1 foot (ft) = 12 inches (in.)	1 square yard (sq yd) = 9 square feet (sq ft) = 1,296 square inches (sq in.) 1 square foot (sq ft) = 144 square inches (sq in.)

Units of Weight	Units of Liquid Volume
1 pound (lb) = 16 ounces (oz) 1 ton (T) = 2,000 pounds (lb)	1 gallon (gal) = 4 quarts (qt) 1 quart (qt) = 2 pints (pt) 1 pint (pt) = 2 cups (c) 1 cup (c) = 8 fluid ounces (fl oz) 1 fluid ounce (fl oz) = 2 tablespoons (tbs) 1 tablespoon (tbs) = 3 teaspoons (tsp)

Units of Time

1 millennium = 10 centuries (cent) = 100 decades = 1,000 years (yr) 1 century (cent) = 10 decades = 100 years (yr) 1 year (yr) = 12 months (mo) = 52 weeks (wk) plus 1 or 2 days = 365 or 366 days	1 month (mo) = 28, 29, 30, or 31 days (d) 1 week (wk) = 7 days (d) 1 day (d) = 24 hours (hr) 1 hour (hr) = 60 minutes (min) 1 minute (min) = 60 seconds (sec)

Data Sources

The lists below give many of the sources for the data in this section. The information on some pages came from just a few sources. On other pages the information came from many sources. Often the information came from one or more sources and then was checked using different sources.

If you look up the data yourself, you might find different information. Data may change as experts discover or collect new information. Or you may find the results of different sets of data from those shown here.

Normal Spring High and Low Temperatures (in °F)

ncdc.noaa.gov

Animal Clutches

animaldiversity.ummz.umich.edu
animals.sandiegozoo.org
www.arkive.org
Cleveland Metroparks
conserveturtles.org
www.fws.gov
National Geographic
nationalzoo.si.edu
wec.ufl.edu

Masses of Sports Balls

Croquet Association
International Federation of Association Football
International Federation of Volleyball
International Swimming Federation
International Table Tennis Federation
Major League Baseball
Marylebone Cricket Club
United States Bowling Congress
United States Golf Association
United States Olympic Committee
United States Tennis Association
USA Basketball
World Squash Federation

U.S. Road Mileage Map

Google Maps

Information about North American Birds

www.allaboutbirds.org
www.iucnredlist.org

Heights of 8-Year-Old Children

www.cdc.gov

Lengths of Day for World Cities

esrl.noaa.gov

Sunrise and Sunset Data for June 21, 2016

esrl.noaa.gov

Train Schedule and Airline Schedule

American Airlines
Regional Transportation Authority

Number of Words Children Know

www.cdi-clex.org

Letter Frequencies

www.math.cornell.edu

How Much Would You Weigh on the Moon?

exploratorium.edu

Physical Fitness Standards

www.cdc.gov
www.presidentschallenge.org

Groveb/E+/Getty Images

About Calculators

Since kindergarten, you have used calculators to do mathematics. You use them to help you learn to count. You use them to operate with numbers. You use them in some of the games you play. Calculators can be helpful tools if you know when and how to use them.

When you use a calculator, always make an estimate of the computation to check that the answer in the display makes sense.

Not all calculators are alike. Some calculators are on computers, tablets, or phones. Here is one type of calculator you may use in class.

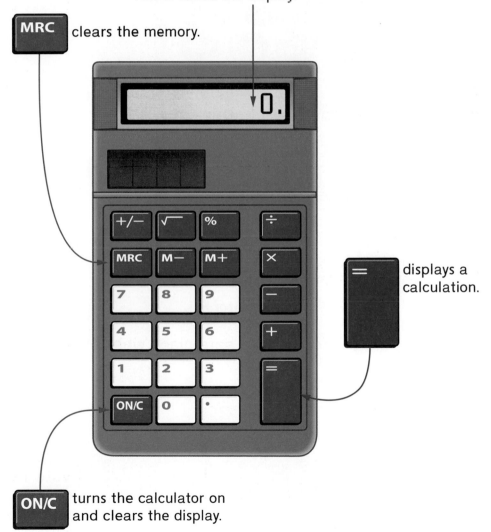

This is called the display.

MRC clears the memory.

= displays a calculation.

ON/C turns the calculator on and clears the display.

Calculator A

Here is another type of calculator you may use.

This is called the display.

C clears the display.

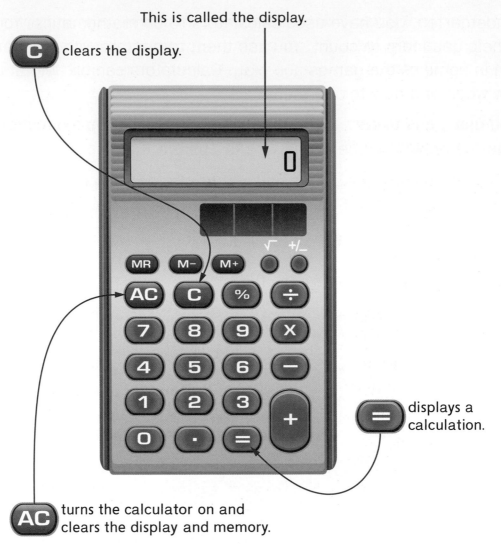

= displays a calculation.

AC turns the calculator on and clears the display and memory.

Calculator B

Take care of your calculator. Keep it in a safe place where you can find it when you need it.

Note Both Calculator A and Calculator B turn off automatically. So, neither calculator has an "Off" key.

Calculating with Basic Operations

A calculator is a tool you can use to add, subtract, multiply, and divide. When you press a key, you enter a direction for your calculator to follow. In *Everyday Mathematics,* most keys are shown in boxes such as (AC), [ON/C], and (=).

A simple direction is to turn on the calculator. Another simple direction is to clear the display. When it is cleared, [0.] is in the display.

Calculator A	
Key	Purpose
[ON/C]	Turn the display on.
[ON/C] [ON/C]	Clear the display and memory.
[ON/C]	Clear only the display.

Calculator B	
Key	Purpose
(AC)	Turn the display on.
(AC)	Clear the display and memory.
(C)	Clear only the display.

The set of keys you press on a calculator is called a *key sequence*. Below you can see key sequences to add, subtract, multiply, and divide numbers. Remember to clear your calculator before starting a new problem.

Operation		Problem	Key Sequence	Display
Add	[+]	23 + 19	23 [+] 19 (=)	42.
Subtract	[−]	42 − 19	42 [−] 19 (=)	23.
Multiply	[×]	6 × 14	6 [×] 14 (=)	84.
Divide	[÷]	84 ÷ 14	84 [÷] 14 (=)	6.

Ask yourself if the answer in the display makes sense by making an estimate. If it does not, then you may have made a mistake entering the key sequence and you should try the problem again. The calculator can only follow the directions you enter and cannot do the thinking for you.

Skip Counting on a Calculator

You can enter a key sequence to use your calculator to skip count up or back. You need to tell the calculator four things:

- what number to count by
- whether to count up or down
- what the starting number is
- when to count

The order of the steps you enter depends on your calculator. The example below shows how to use Calculator A.

Example

Starting at 1, count up by 2s.

Calculator A			
	Purpose	Key Sequence	Display
	Clear the memory and display.	ON/C ON/C	0.
	Tell the calculator to start at 1 and count up.	1 +	1.
	Tell the calculator to count by 2s and do the first count.	2 =	3.
	Tell the calculator to count again.	=	5.
	Keep counting by pressing =.	=	7.

To count back by 2s, begin with the starting number followed by −.

The example below shows how to use Calculator B to skip count.

Example

Starting at 1, count up by 2s.

Calculator B			
	Purpose	Key Sequence	Display
	Clear the memory and display.	AC	0.
	Tell the calculator to count up by 2.	2 + +	K 2.▪
	Tell the calculator to start at 1 and do the fist count.	1 =	K 3.▪
	Tell the calculator to count again.	=	K 5.▪
	Keep counting by pressing =.	=	K 7.▪

To count *back* by 2s, begin with ② ⊖ ⊖.

How can you tell the calculator you use to skip count? If the instructions on this page and the one before do not work for your calculator, read the instruction manual that came with your calculator or ask an adult to help you find instructions online.

Note The "K" on Calculator B's display means "constant." It means the calculator knows the count-by number and the direction (up or back).

Check Your Understanding

Use your calculator to find the answer. Before you press =, make an estimate to check if your answer is reasonable.

1. 331 + 456 = ?

2. 310 − 190 = ?

3. 66 × 5 = ?

4. 342 ÷ 9 = ?

Use your calculator to skip count. Record the counts.

5. Starting at 5, count up by 3s. Stop at 20.

6. Starting at 10, count back by 2s. Stop at 0.

Check your answers in the Answer Key.

Order of Operations on a Calculator

Mathematicians agree to follow certain rules in mathematics so everyone using the same **operations** in a number sentence gets the same answer.

	Rules for the Order of Operations
1.	Do the operations inside parentheses first. Follow Rules 2 and 3 when computing inside parentheses.
2.	Multiply or divide, in order, from left to right.
3.	Add or subtract, in order, from left to right.

Note If multiplication and division (or addition and subtraction) are in the same number sentence, they have equal priority. Perform them in order from left to right. For example, the division comes first in the problem $8 \div 2 \times 5 = ?$, so divide first and then multiply.

$$8 \div 2 \times 5 = ?$$
$$4 \times 5 \quad = 20$$

Since many calculators, such as Calculator A and Calculator B, do not have parentheses and some only do operations as you enter them, you must know these rules. If your calculator does not follow the order of operations, you must follow the rules when you enter a key sequence.

Example

Solve. $3 + 4 \times 5 = ?$ Remember to follow the rules of the order of operations.

The order of operations starts with parentheses. There are no parentheses, so go to the next rule: multiply or divide.

$3 + \mathbf{4 \times 5}$
$4 \times 5 = \mathbf{20}$

Then add (or subtract) from left to right.

$3 + \mathbf{20} = 23$

Other calculators do use the order of operations. Here's how to find out if a calculator uses the order of operations:

1. Solve the problem $1 + 2 \times 3$. Following the rules, you know that you must multiply first: $2 \times 3 = 6$. Then add 6 to 1: $1 + 6 = 7$.

2. Now enter the key sequence ⌑1⌑ ⌑+⌑ ⌑2⌑ ⌑×⌑ ⌑3⌑ ⌑=⌑. If the display shows "7," the calculator follows the order of operations. If the display shows "9," it does not.

2-dimensional (2-D)
Having *area* but not volume. A 2-dimensional *surface* can be flat like a piece of paper or curved like a dome.

3-dimensional (3-D)
Having length, width, and thickness. Solid objects that take up space, such as balls, rocks, boxes, and books, are 3-dimensional.

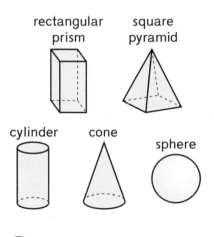

rectangular prism   square pyramid

cylinder   cone   sphere

A

A.M. An abbreviation that means "before noon." It refers to the period between midnight (12 A.M.) and noon (12 P.M.).

accurate As correct as possible for the situation.

addend Any one of a set of numbers that are added. For example, in $5 + 3 + 1 = 9$, the addends are 5, 3, and 1.

adding a group A multiplication fact strategy that involves adding one more group onto a helper fact. For example, 6×8 can be solved by starting with 5 groups of 8 ($5 \times 8 = 40$), then adding one more group of 8 to get 48.

addition fact Two whole numbers from 0 through 10 and their sum, such as $9 + 7 = 16$.

algebra The branch of mathematics that uses letters and symbols to stand for *unknowns*. Algebra is used to *model patterns*, numerical relationships, and real-world situations.

algorithm A set of step-by step instructions for doing something, such as carrying out a computation or solving a problem.

analog clock A clock that shows the time by the positions of the hour and minute hands.

angle A figure that is formed by two line segments that have the same endpoint.

approximate Close to exact. It is sometimes not possible to get an exact answer, but it is important to be close to the exact answer.

area The amount of *surface* inside a *2-dimensional* shape. The measure of the area is how many units, such as square inches or square centimeters, cover the surface.

40 square units   about 21 square units

1 square centimeter   1 square inch

area model A *model* for multiplication problems in which the length and width of a *rectangle* represent the *factors*, and the *area* of the rectangle represents the *product*.

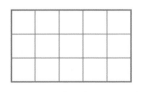

Area model for $3 * 5 = 15$

array An arrangement of objects into *rows* and *columns* that form a rectangle. All rows and columns must be filled. Each row has the same number of objects. And each column has the same number of objects.

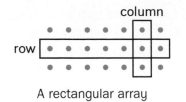

A rectangular array

B

ballpark estimate A rough *estimate* to help you solve a problem, check an answer, or when an exact answser cannot be found.

bar graph A graph that uses horizontal or vertical bars to represent *data.*

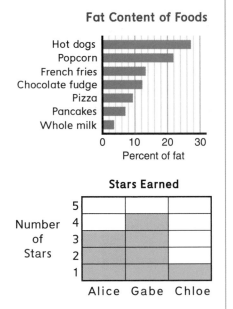

Fat Content of Foods

Stars Earned

base The *side* of a polygon or *face* of a polyhedron from which the *height* is measured.

Bases are shown in red.

base ten Our system for writing numbers that uses 10 symbols called *digits.* The digits are 0, 1, 2, 3, 4, 5, 6, 7, 8, and 9. You can write any number using only these 10 digits. Each digit has a value that depends on its place in the number.

basic facts The *addition facts* (whole-number addends of 10 or less) and their related subtraction facts, and the multiplication facts (whole number factors of 10 or less) and their related division facts. Facts are organized into *fact families.*

benchmark A well-known number or measure that can be used to check whether other numbers, measures, or estimates make sense. For example, a benchmark for length is that the width of a man's thumb is about one inch. The numbers 0, $\frac{1}{2}$, 1, 1 $\frac{1}{2}$ may be useful benchmarks for fractions.

break apart A multiplication fact strategy where you decompose a factor into smaller numbers. For example, 7 × 6 can be broken apart into 2 × 6 and 5 × 6 to make solving easier.

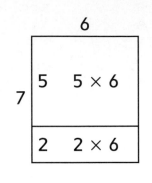

C

change diagram A diagram used in *Everyday Mathematics* to represent situations in which quantities are increased or decreased.

circle A 2-dimensional, closed, curved path whose points are all the same distance from a center point.

Circle

close-but-easier numbers Numbers that are close to the original numbers in the problem, but easier for solving problems. For example, to estimate 494 + 78, you might use the close-but-easier numbers 480 and 80.

column A vertical ("up and down") arrangement of object in an *array.*

column

column addition A method for adding numbers in which the *place-value* columns are separated. The digits in each column are added, and then trades are made until each column has only one digit.

100s	10s	1s
2	4	8
+ 1	8	7
3	12	15
3	13	5
4	3	5

248 + 187 = 435

combinations of 10 Pairs of whole numbers from 0 to 10 that add to 10. For example, 4 + 6 and 0 + 10 are both combinations of 10.

comparison diagram A diagram used in *Everyday Mathematics* to represent situations in which two quantities are compared.

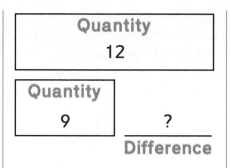

Quantity
12

Quantity	
9	?

Difference

compose To make a number or shape by putting together smaller numbers or shapes. For example, you can compose a 10 by putting together ten 1s: 1 + 1 + 1 + 1 + 1 + 1 + 1 + 1 + 1 + 1 = 10. You can compose a pentagon by putting together an equilateral triangle and a square.

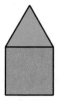

A composed pentagon

composite unit A *unit* of measure made up of smaller units. For example, a foot is a composite unit of 12 inches, and a row of unit squares can be used to measure area.

3 rows of 5 square units each have a total area of 15 square units.

conjecture A statement that is thought to be true based on information or mathematical thinking.

counting numbers The numbers used to count things: 1, 2, 3, 4, and so on.

counting-up subtraction A subtraction strategy in which you count up from the smaller to the larger number to find the *difference.* For example, to solve 16 − 9, count up from 9 to 16.

cube (1) A polyhedron with 6 square *faces.* A cube has 8 *vertices* and 12 *edges.* (2) The smallest base-10 block called a centimeter cube.

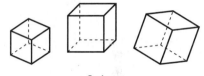

Cubes

D

data Information that is gathered by counting, measuring, asking questions, or observing.

decompose To separate a number or shape into smaller numbers or shapes. For example, you can decompose 14 into 1 ten and 4 ones. You can decompose a square into two triangles.

denominator The number below the line in a *fraction.* For example, in $\frac{3}{4}$, 4 is the denominator.

difference The result of subtracting one number from another.

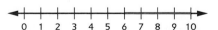

digit One of the number symbols 0, 1, 2, 3, 4, 5, 6, 7, 8, and 9 in the standard, *base-ten* system.

digital clock A clock that shows the time with numbers of hours and minutes, usually separated by a colon. For example, this digital clock shows three o'clock.

A digital clock

dividend The number in division that is being divided. For example, in 35 ÷ 5 = 7, the dividend is 35.

divisor In division, the number that divides another number. For example, in 35 ÷ 5 = 7, the divisor is 5.

double (1) Two times an amount; an amount added to itself. (2) A multiplication strategy in which you multiply the product of a helper fact by 2.

edge Any *side* of a *polyhedron's faces.*

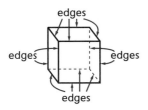

efficient strategy A method that can be applied easily and quickly. For example, adding a group and doubling are usually efficient *strategies* for solving multiplication facts.

elapsed time An amount of time that has passed. For example, between 12:45 P.M. and 1:30 P.M., 45 minutes have elapsed.

equal See *equivalent.*

equal groups Collections or groups of things that all contain the same number of things. For example, boxes that each contain 100 clips are equal groups. And rows of chairs with 6 chairs per row are equal groups.

equal parts *Equivalent* parts of a *whole.* For example, dividing a pizza into 4 equal parts means each part is $\frac{1}{4}$ of the pizza and is equal in size to each of the other 3 parts.

equation A *number sentence* that contains an equal sign. For example, 15 = 10 + 5 is an equation.

equivalent *Equal* in value but possibly in a different form. For example, 3, 1 + 1 + 1, and $\frac{6}{2}$ are all equivalent.

equivalent fractions *Fractions* that name the same number. For example, $\frac{1}{2}$ and $\frac{4}{8}$ are equivalent fractions.

equivalent names Different ways to name the same number. For example, 2 + 6, 12 − 4, 2 × 4, 16 ÷ 2, 5 + 1 + 2, VIII, eight, and are equivalent names for 8.

estimate An answer close to an exact answer. To estimate means to give an answer that should be close to an exact answer.

even number A *counting number* that can be divided by 2 with no remainder. The even numbers are 2, 4, 6, 8, and so on.

expand-and-trade subtraction A way to subtract that uses expanded forms to help you make *place-value* exchanges.

expanded form A way of writing a number as the *sum* of the values of each *digit.* For example, in expanded form, 356 is written 300 + 50 + 6. Compare *standard form.*

extended facts Basic facts changed to involve *multiples* of 10, 100, and so on. For example, 30 + 70 = 100, 40 ∗ 5 = 200, and 560 / 7 = 80 are extended facts.

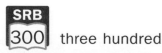

F

face A flat *surface* on the outside of a solid.

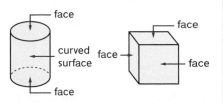

fact extension Arithmetic with larger numbers you can do knowing *basic facts*. For example, knowing the addition fact $5 + 8 = 13$ makes it easier to solve fact extensions such as $50 + 80 = ?$ or $1300 - ? = 500$.

fact family (1) A set of related addition and subtraction facts. For example, $5 + 6 = 11$, $6 + 5 = 11$, $11 - 5 = 6$, and $11 - 6 = 5$ are a fact family. (2) A set of related multiplication and division facts. For example, $5 \times 7 = 35$, $7 \times 5 = 35$, $35 \div 5 = 7$, and $35 \div 7 = 5$ are a fact family.

Fact Triangle Cards with a triangle shape that show *fact families*. Fact Triangles are used like flash cards to help you memorize basic addition, subtraction, multiplication, and division facts.

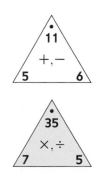

factor (1) Any of the numbers that are multiplied to find a *product*. For example, in the problem $4 \times 7 = 28$, 28 is the product, and 4 and 7 are the factors. (2) A number that divides another number evenly. For example, 8 is a factor of 24 because $24 \div 8 = 3$, with no remainder.

factor pair Two *factors* of a *counting number* whose *product* is the number. A number may have more than one factor pair. For example, the factor pairs for 18 are 1 and 18, 2 and 9, and 3 and 6.

facts table A chart with rows and columns that shows all of the basic addition and subtraction facts, or all of the basic multiplication and division facts.

false number sentence A *number sentence* that is not true. For example, $8 = 5 + 5$ is a false number sentence.

fraction A number in the form $\frac{a}{b}$ or a/b. Fractions can be used to name part of a whole or part of a collection. The number a is called the *numerator*. The number b is called the *denominator* and cannot be 0.

Fraction Circle Pieces A set of colored circles each divided into equal-size slices, used to represent *fractions*.

Frames and Arrows A diagram used in *Everyday Mathematics* to show a number pattern or sequence.

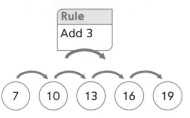

front-end estimation An estimation method that keeps only the left-most digit in the numbers and puts 0s in for all others. For example, the front-end *estimate* for $4,560 + 5,345$ is $4,000 + 5,000 = 9,000$.

function machine An imaginary machine that uses a rule to pair input numbers put in (inputs) with numbers put out (outputs). Each input is paired with exactly one output. Function machines are used in "What's My Rule" problems.

G

geoboard A small board with posts that are usually equally spaced in a rectangular array.

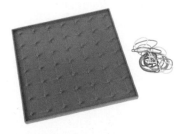

Geoboard and rubber bands

Glossary

geometric solid A *3-dimensional* shape, such as a *prism*, *pyramid,* cylinder, cone, or sphere. Despite its name, a geometric solid is hollow; it does not contain the points in its interior.

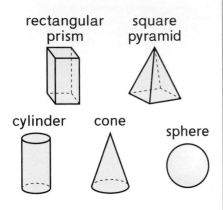

rectangular prism   square pyramid

cylinder   cone   sphere

going through 10 A subtraction fact strategy in which you combine *differences* between each number and 10. For example, to solve 16 − 9, solve 16 − 10 = 6 and 10 − 9 = 1, then add 6 + 1 = 7.

grouping addends An addition strategy in which you add three or more numbers in an order that makes the addition simpler, such as adding a combination of 10 or a doubles fact first.

 **H**

height (1) The length of the shortest line segment joining a corner of a shape to the line containing the base opposite it. (2) The line segment itself.

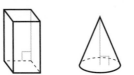

Heights shown in blue

helper facts Well-known facts you can use to solve other facts. For example, if you know 2 × 7 = 14, then you can add a group of 7, 14 + 7 = 21 to find 3 × 7 = 21.

I

inequality A *number sentence* with >, <, ≥, or ≤. For example, the sentence 8 < 15 is an inequality.

iterate units To repeat a *unit* without gaps or overlaps in order to measure. For example, you can cover a surface by iterating or tiling a unit to measure *area.*

K

kilogram A metric unit of *mass* equal to 1,000 grams. A bottle of water is usually 1 kilogram.

kite A 4-sided polygon with two pairs of equal length *sides.* The equal sides are next to each other. The four sides can all have the same length. (So a rhombus is a kite.)

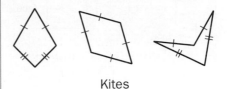

Kites

L

length The distance between two points along a path.

length of day The amount of time that passes between sunrise and sunset.

line A straight path that goes on forever in both directions.

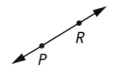

line *PR* or *RP*

line plot A sketch of *data* that uses Xs or other marks above a number line to show how many times each value appears in the set of data.

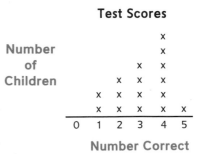

Test Scores

			x		
Number			x		
of		x	x		
Children		x	x	x	
	x	x	x	x	
	x	x	x	x	x
0	1	2	3	4	5

Number Correct

line segment A straight path joining two points. The two points are called endpoints of the segment.

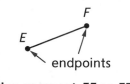

line segment *EF* or *FE*

liquid volume An amount of liquid measured in units, such as liters and gallons. Units of liquid *volume* are frequently used to measure how much a volume container can hold.

long In *Everyday Mathematics*, the base-10 block that is equivalent to ten 1-centimeter cubes.

making 10 An *addition fact* strategy in which you break apart a number so you can get a sum of 10. For example, you can solve 9 + 5 by breaking 5 into 1 and 4, then thinking 9 + 1 = 10 and 4 more makes 14.

mass A measure of how much matter is in object. Mass is often measured in grams or kilograms.

mathematical argument An explanation of why a claim is true or false using words, pictures, symbols, or other representations. For example, if you claim that $\frac{2}{5} > \frac{4}{7}$ is not true, you say that $\frac{2}{5}$ is less than $\frac{1}{2}$ and $\frac{4}{7}$ is more than $\frac{1}{2}$, so $\frac{2}{5}$ cannot be more than $\frac{4}{7}$.

mathematical practices Ways of working with mathematics. Mathematical practices are habits or actions that help people use mathematics to solve problems.

measurement scale The spacing of the marks on a measuring device. The scales on this ruler are 1 millimeter on the left side and $\frac{1}{16}$ inch on the right side.

Scale of a number line

metric system A measurement system based on multiples of 10. The metric system is used by scientists and people in most countries in the world except the United States.

model A representation of a real-world object or situation. Number sentences, diagrams, and pictures can be models.

multiple of a number *n* A *product* of *n* and a *counting number*. For example, the multiples of 7 are 7, 14, 21, 28, and so on.

multiplication square Same as *square number*.

multiplication/division diagram A diagram used for problems in which there are several equal groups. The diagram has three parts: a number of groups, a number in each group, and a total number.

N

name-collection box In *Everyday Mathematics*, a place to write *equivalent names* for a number.

50	
100 ÷ 2	5 × 10
10 + 10 + 10 + 10 + 10	
1 more than 49	25 + 25
fifty	*cincuenta*

near doubles An addition fact strategy in which you add or subtract from a doubles fact you know to solve another fact. For example, you can solve 7 + 8 by thinking 7 + 7 = 14, 14 + 1 = 15.

near squares A special kind of *adding or subtracting a group* where the *helper fact* is a multiplication *square*. For example, to solve 7 × 8, you may use 8 × 8 as a helper fact, and then subtract a group of 8.

negative numbers A number that is less than zero; a number to the left of zero on a horizontal *number line* or below zero on a vertical number line. The symbol − may be used to write a negative number. For example, "negative 5" is usually written as −5.

number line A *line* with numbers marked in order on it.

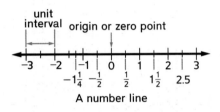

A number line

number model A group of numbers and symbols that shows how a *number story* can be solved. For example, 10 − 6 = 4 and 10 − 6 are each number models for the following story: *I had 10 bookmarks. I gave 6 away. How many did I have left?*

number sentence Two groups of mathematical symbols connected by a *relation symbol* (=, >, <, ≠). Mathematical symbols on each side of the number sentence include numbers and/or operation symbols (+, −, ×, or ÷). For example, 2 × 5 = 10 and 6 + 3 + 4 = 2 + 11 are number sentences.

number story A story with a problem that can be solved using arithmetic.

numerator The number above the line in a *fraction.* For example, in $\frac{3}{4}$, 3 is the numerator.

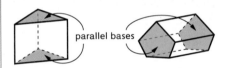

odd number A *counting number* that cannot be exactly divided by 2. When an odd number is divided by 2, there is a remainder of 1. The odd numbers are 1, 3, 5, and so on.

open number line A line on which you can mark points or numbers that are useful for solving problems.

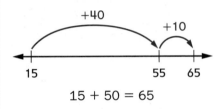

15 + 50 = 65

operation An action performed on numbers to produce other numbers. Addition, subtraction, multiplication, and division are the four basic arithmetic operations.

operation symbol A symbol used to stand for a mathematical operation. Common operation symbols are +, −, ×, ÷.

order of operations Rules that tell in what order to perform operations in arithmetic.

P

P.M. An abbreviation that means "after noon." It refers to the period between noon (12 P.M.) and midnight (12 A.M.).

pan balance A tool used to weigh objects or compare *weights.*

parallel Two *lines* are parallel if they are always the same distance apart, and never meet or cross each other, no matter how far they are extended. Line segments are parallel if they are parts of lines that are parallel. The bases of a prism are parallel.

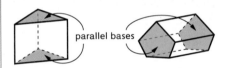

parallel bases

parallelogram A 4-sided polygon whose opposite sides are *parallel.* The opposite sides of a parallelogram are also the same length. And the opposite angles in a parallelogram have the same measure. All parallelograms are *trapezoids.*

Parallelogram

parentheses Grouping symbols, ( ), used to tell what should be calculated first.

partial-sums addition A way to add in which *sums* are computed for each place (ones, tens, hundreds, and so on) separately. The partial-sums are then added to give the final answer.

partition In geometry, to divide a shape into smaller shapes. For example, shapes can be partitioned into equal shares to represent *fractions*. Partitioning can also be used to find length, *area,* or *volume.*

parts-and-total diagram A diagram you can use to represent number stories that combine two or more quantities to make a total quantity.

Total	
13	
Part	**Part**
8	?

Parts-and-total diagram for
13 = 8 + ?

pattern Shapes or numbers ordered by a rule so that what comes next can be predicted.

per "For each" or "in each." For example, "three tickets per student" means "three tickets for each student."

perimeter The distance around the boundary of a shape. The perimeter of this triangle is 15 ft.

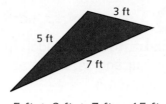

5 ft + 3 ft + 7 ft = 15 ft

picture graph A graph that uses pictures or symbols to represent *data.* The key for a picture graph tells what each picture or symbol is worth.

Trees Planted in Park

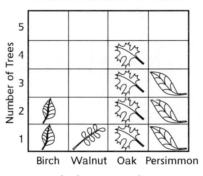

A picture graph

place value A system for writing numbers in which the value of a *digit* depends on its place in the number.

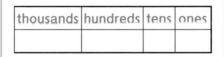

thousands	hundreds	tens	ones

A place-value chart

plot To draw on a number line or graph. The points plotted may come from *data.*

point An exact location in space.

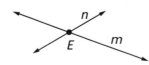

Lines *m* and *n* intersect at point *E*.

polygon A 2-dimensional shape that is made up of *line segments* joined end to end. The line segments make one closed path and may not cross.

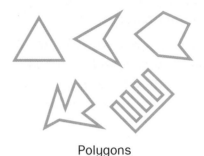

Polygons

polyhedron A solid whose surfaces (called *faces*) are all flat and formed by *polygons.* The faces may meet but not cross. A polyhedron does not have any curved surfaces.

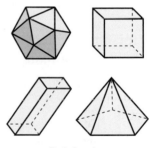

Polyhedra

positive numbers Numbers that are greater than zero. Positive numbers are usually written to the right of zero on a horizontal number line or above zero on a vertical number line. A positive number may be written using the + symbol, but is usually written without it. For example, +10 = 10.

precise Exact. The smaller the *unit* used in measuring, the more precise the measurement is. For example, a measurement to the nearest inch is more precise than a measurement to the nearest foot.

prism A polyhedron that has two *parallel bases* that are formed by polygons with the same size and shape. The other *faces* connect the bases and are shaped like *parallelograms*. These other faces are often rectangles. Prisms take their names from the shape of their bases.

Triangular prism   Rectangular prism   Hexagonal prism

product The result of multiplying two or more numbers, called *factors*. For example, in 4 * 3 = 12, the product is 12.

pyramid A *polyhedron* in which one *face*, the *base*, may have any polygon shape. All of the other faces are triangular and come together at the apex. A pyramid takes its name from the shape of its base.

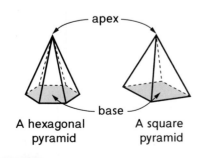

apex

base

A hexagonal pyramid   A square pyramid

quadrangle A polygon that has four angles. Same as *quadrilateral*.

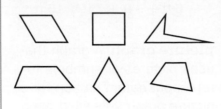

quadrilateral A *polygon* that has four sides. Same as *quadrangle*.

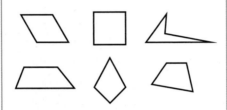

quantity A number with a unit, usually a measurement or count.

quotient The result of dividing one number by another number. For example, in 35 ÷ 5 = 7, the quotient is 7.

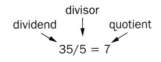

divisor

dividend   quotient

35/5 = 7

rectangle A *parallelogram* whose corners are all *right angles*.

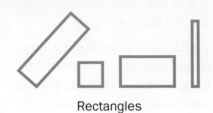

Rectangles

rectangular prism A *prism* with rectangular *bases*. The four *faces* that are not bases are either *rectangles* or other *parallelograms*. A rectangular prism may model a shoebox.

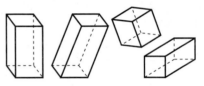

Rectangular prisms

rectilinear figure A polygon with a right angle at each vertex.

Rectilinear figures

regular polygon A *polygon* whose sides are all the same length and whose interior *angles* are all the same measure. For example, a *square* is a regular polygon.

relation symbol A symbol that shows a relationship between two quantities. For example, three is less than five can be written as 3 < 5. Some relationship symbols are <, >, and =.

remainder The amount left over when things are divided or shared equally. Sometimes there is no remainder.

represent To show or stand for something. For example, numbers can be represented using base-10 blocks, spoken words, or written numerals.

rhombus A *quadrilateral* whose four sides are all the same length. All rhombuses are *parallelograms* and *kites.* Every square is a rhombus, but not all rhombuses are squares.

Rhombuses

right angle An *angle* whose sides form a square corner.

round To change a number slightly to make it easier to work with. Often, numbers are rounded to the nearest 10, 100, 1,000, and so on. For example, 864 rounded to the nearest hundred is 900.

row A horizontal ("side to side") arrangement of object in an *array.*

 row

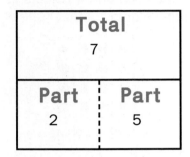

scale (1) A comparison between the number of units in a picture or model and the actual number of units. A picture graph may show 1 smiley face to stand for 10 people. (2) See *scale of a number line* or *measurement scale.* (3) A tool for measuring *weight* or *mass.*

scale of a number line The spacing of the marks on a *number line.* The scale of the number line below is halves.

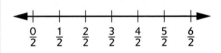

second (s or sec) (1) The basic *unit* of time. Minutes, hours, and days are based on seconds. (2) An ordinal number in the sequence first, second, third, . . .

sequence A list of numbers, often created by a rule that can be used to put more numbers in the list. Frames-and-Arrows diagrams can represent sequences.

side (1) One of the *line segments* of a *polygon.* (2) One of the line segments that make up an *angle.* (3) One of the *faces* of a solid figure.

situation diagram In *Everyday Mathematics*, a diagram used to organize information in a problem situation.

Total	
7	
Part	**Part**
2	5

Suzie has 2 pink balloons and 5 yellow balloons. She has 7 balloons in all.

solid See geometric solid.

solution of an open sentence A value that makes an open sentence true when it is substituted for the variable. For example, 7 is a solution of $5 + n = 12$.

square A *rectangle* whose sides are all the same length. A rectangle that is also a *rhombus.*

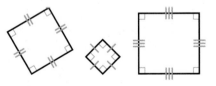

Squares

square array A rectangular *array* with the same number of rows as columns. For example, 16 objects will form a square array with 4 objects in each row and 4 objects in each column.

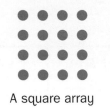

A square array

square number A number that is the *product* of a counting number with itself. For example, 25 is a square number because 25 = 5 * 5. The square numbers are 1, 4, 9, 16, 25, and so on. A square number can be represented by a *square array.*

square unit A *unit* used in measuring *area,* such as a square centimeter or a square foot.

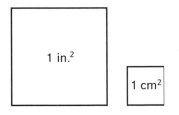

Square units

standard form The most familiar way of representing numbers. In standard form, numbers are written using the *base-ten place-value* system. For example, standard form for three hundred fifty-six is 356. Compare *expanded form*.

standard notation Same as standard form.

standard unit Measurement *units* that are the same size no matter who uses them and when or where they are used.

strategy An approach to a problem that may be general, like "trial and error," or more specific, like "break apart."

subtracting a group A multiplication fact strategy using a *helper fact* with a larger factor. For example, 9 × 6 can be solved by using the helper fact 10 × 6 and subtracting one group of 6 from the product.

sum The result of adding two or more numbers. For example, in 5 + 3 = 8, the sum is 8.

surface A *2-dimensional* layer, such as a filled-in polygon, the top of a body of water, or the outside of a ball.

T

tally chart A chart that uses marks, called tallies, to show how many times each value appears in a set of data.

Number of Pull-Ups	Number of Children
0	HHT /
1	HHT
2	////
3	//
4	
5	///
6	/

temperature A measure of how hot or cold something is.

think addition A subtraction fact strategy in which you think of an *addition fact* in the same fact family. For example, you can solve 10 − 4 = ? by thinking 4 + ? = 10 and knowing that 4 + 6 = 10.

tile (verb) To cover a *surface* completely with shapes without overlaps or gaps. Tiling with same-size squares is a way to measure area.

tool Anything that can be used for performing a task. Calculators, rulers, *situation diagrams,* and number grids are examples of mathematical tools.

trade-first subtraction A subtraction method in which all trade are done before any subtractions are carried out.

trapezoid A 4-sided polygon that has at least one pair of *parallel* sides.

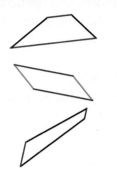

Parallel bases are marked in the same color

triangle A *polygon* that has 3 sides and 3 angles.

triangular number A counting number that can be shown by a triangular arrangement of dots. The triangular numbers are 1, 3, 6, 10, 15, 21, 28, 36, 45, and so on.

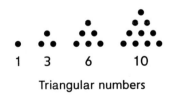

1   3   6   10

Triangular numbers

triangular prism A *prism* whose *bases* are triangles.

Triangular prisms

true number sentence A *number sentence* in which the relation symbol accurately connects the two sides. For example, 15 = 5 + 10 and 25 > 20 + 3 are both true number sentences.

turn-around rule A rule that says that numbers can be added or multiplied in either order. For example, if you know that 6 + 7 = 13, then, by the turn-around rule for addition, you also know that 7 + 6 = 13. If you know that 4 × 5 = 20, then, by the turn-around rule for multiplication, you also know that 5 × 4 = 20.

U

U.S. customary system The measurement system that is used most commonly in the United States.

unit A label used to put a number in context. In measuring length, for example, the inch and the centimeter are units. In a problem about 5 apples, *apple* is the unit.

unit fraction A *fraction* whose *numerator* is 1. For example, $\frac{1}{2}$, $\frac{1}{3}$, $\frac{1}{8}$, and $\frac{1}{20}$ are unit fractions.

unit square A *square* with side lengths of 1.

unknown A quantity whose value is not known. An unknown is sometimes represented by a ____, a ?, or a letter.

V

vertex A point where the *sides* of an *angle,* the sides of a polygon, or the *edges* of a *polyhedron* meet. 2 or more are called vertexes or vertices.

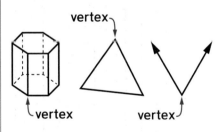

vertex

vertex   vertex

volume The amount of space an object takes up. Volume is often measured in liquid units, such as liters, or cubic units, such as cubic centimeters or cubic inches. The volume or capacity of a container is a measure of how much the container will hold.

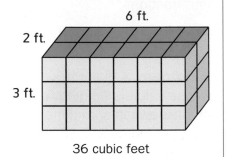

36 cubic feet

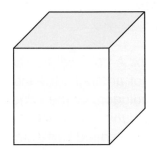

1 cubic inch

1 cubic centimeter

 W

weight A measure of how heavy something is.

"What's My Rule?" A type of problem with "in" numbers, "out" numbers, and a rule that changes the in numbers to the out numbers. Sometimes you have to find the rule. Other times, you use the rule to figure out the in or out numbers.

whole An entire object, collection of objects, or quantity being considered.

Y

yard A U.S. customary unit of length equal to 3 feet, or 36 inches.

Z

zero (1) The number representing no amount. (2) To adjust a scale or balance before use so that it reads 0 when no object is being weighed. You need to zero a scale for *accurate* measurements.

Page 14

1. Tiffany had to show that Martin's conjecture didn't match one of the clues.

2. Martin showed that it fit both clues.

3. Sample answer: My conjecture is 26. The conjecture is true because it fits both clues. 26 erasers can be shown as 5 groups of 5 with 1 left over and 8 groups of 3 with 2 left over.

Page 17

1. Sample answer: For every 3 red flowers, there are 2 yellow flowers, for a total of 5 flowers. Looking at the third row, if I follow this pattern, then for 9 red flowers, I have 6 yellow flowers for a total of 15 flowers.

2. Sample answer: They both show how they add 3 red flowers for every 2 yellow flowers. Martin uses a table, while Tiffany uses a drawing.

Page 22

No. Noah did not read 500 pages in 3 weeks. I found my answer by using close-but-easier numbers to estimate the number of pages he read in 3 weeks. 97 is close to 100 pages, 211 is close to 200 pages, and 143 is close to 150 pages. $100 + 200 + 150 = 450$ pages. 450 pages is fewer than 500 pages.

Page 26

1. Sample answer: I see 24 dots: 3 groups of 4 across the top and 3 groups of 4 across the bottom. $12 + 12 = 24$ dots

2. Sample answer: My strategy is similar to Bella's because we both saw groups of dots that we multiplied.

Page 31

1. 5 pennies

2. 14 boys

3. 4 and 3

4. 14¢

Page 43

1. Sample answers: $4 \times 5 = 20$; $5 \times 4 = 20$; $20 \div 4 = 5$; $20 \div 5 = 4$

2. Sample answers: $2 \times 9 = 18$; $9 \times 2 = 18$; $18 \div 2 = 9$; $18 \div 9 = 2$

3. 8 plants per row; Sample answer:

```
• • • • • • • •
• • • • • • • •
• • • • • • • •
• • • • • • • •   32 ÷ 4 = 8
```

Page 48

1. 27; Sample answer: I used the helper fact $2 \times 9 = 18$. Then I added one more group of 9. Since $18 + 9 = 27$, I know $3 \times 9 = 27$.

2. 48

3. 72; Sample answer: I used the helper fact $10 \times 8 = 80$, then I subtracted a group of 8. Since $80 - 8 = 72$, I know $9 \times 8 = 72$.

4. 32

Page 50

1. 36

2. 48; Sample answer: I used $3 \times 8 = 24$. Then I doubled the product: $24 + 24 = 48$. So, $6 \times 8 = 48$.

3. 30; Sample answer: I used the square fact $5 \times 5 = 25$. Then I added another group of 5. Since $25 + 5 = 30$, I know $5 \times 6 = 30$.

Page 52

1. 79

2. 0

3. 140; Sample answer: $(7 \times 4) \times 5 = 28 \times 5 = 140$; $7 \times (4 \times 5) = 7 \times 20 = 140$. Multiplying $7 \times (4 \times 5)$ was easier than $(7 \times 4) \times 5$ because I could do the extended fact 7×20 in my head more easily than 28×5.

4. 48; Sample answer: Break 6 into 5 and 1, so $5 \times 8 + 1 \times 8 = 48$.

Page 53

1. $3 \times 9 = 27$; $9 \times 3 = 27$; $27 \div 3 = 9$; $27 \div 9 = 3$

2.

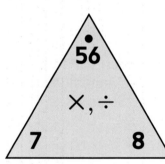

3. Answers vary.

Page 56

Products of 10s facts are also products of 5s facts. For 5s facts, the digit in the ones place is either 0 or 5. For 10s facts, the digit in the ones place is always 0. The products of 5s facts are found in the 5s column and the 10s column. The products of 10s facts are only found in the 10s column.

Page 58

1. $7 \times 4 = 28$; 280

2. $8 \times 3 = 24$; 2,400

3. $36 \div 4 = 9$; 90

4. $30 \div 6 = 5$; 50

Page 70

1. $L = 8$

2. $j = 0$

3. $B = 1$

Page 78

1. Miss Fry needs 6 more canvases. Sample answer:

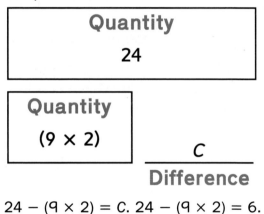

$24 - (9 \times 2) = C$. $24 - (9 \times 2) = 6$.

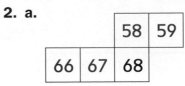

Page 88

1. measure

2. code

3. location

4. compare

5. measure

6. location

7. measure

8. count

9. compare

10. location

11. location

12. count

Page 92

1. a. 30

   b. 25

   c. 35

2. a.

		58	59
66	67	68	

b.

213		215	216
	224	225	

c.

	31		33	
40		42		44

Page 97

1. 9

2.

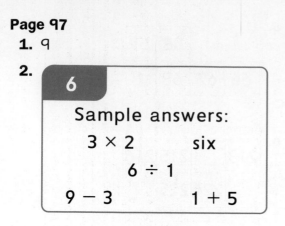

> **6**
>
> Sample answers:
>
> 3 × 2   six
>
>   6 ÷ 1
>
> 9 − 3   1 + 5

3. 10 + 10 + 6 and 9 × 3 + 3 do not belong. Sample explanation: 10 + 10 + 6 = 26, not 36. It's not an equivalent name for 36, so it does not belong in the box.

Page 102

Yes, since $49 + $23 + $24 is a little less than $50 + $25 + $25 = $100.

Page 107

1. **a.** 370 books; 210 + 160 = 370

   b. 400 books; 200 + 200 = 400

   c. 375 books

   d. Yes, both estimates are reasonable. Sample answer: I prefer the estimate rounded to the nearest 10 because it is closer to the exact answer.

Page 115

1. 100

2. 900

3. 30

4. 700

5. 32

6. 70

Page 118

1. 133; Sample estimate: 40 + 100 = 140

2. 386; Sample estimate: 160 + 230 = 390

3. 848; Sample estimate: 500 + 360 = 860

4. 241; Sample estimate: 150 + 90 = 240

Page 119

1. 47; Sample estimate: 90 − 40 = 50

2. 384; Sample estimate: 850 − 450 = 400

3. 372; Sample estimate: 500 − 150 = 350

Page 123

1. 57; Sample estimate:
   $90 - 30 = 60$

2. 172; Sample estimate:
   $240 - 70 = 170$

3. 355; Sample estimate:
   $740 - 400 = 340$

4. 181; Sample estimate:
   $360 - 180 = 180$

Page 138

1. Sample answers: $\frac{3}{8}$, 3-eighths, or three-eighths

2. Sample answers: $\frac{2}{3}$, 2-thirds, or two-thirds

3. 1-fourth of a red fraction circle piece is bigger than 1-fourth of a pink fraction circle piece because the whole red fraction circle piece is bigger than the whole pink fraction circle piece.

4. Sample answers: $\frac{16}{6}$ or $2\frac{4}{6}$

Page 141

$\frac{2}{3}$

Page 143

$\frac{5}{4}$ or $1\frac{1}{4}$

Page 144

1.
   0   $\frac{1}{2}$ 1

2.
   0   $\frac{3}{4}$ 1

3.
   0   1 $\frac{4}{3}$ 2

Page 151

1. equivalent

2. not equivalent

Page 162

1. $\frac{7}{2}$ inches or $3\frac{1}{2}$ inches

2. $\frac{1}{4}$ cup

3. $\frac{4}{6}$ loaf or $\frac{1}{2}$ plus $\frac{1}{6}$ loaf; Sample drawing:

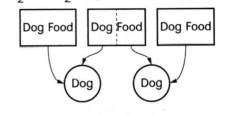

4. $\frac{3}{2}$, or $1\frac{1}{2}$ bags; Sample drawing:

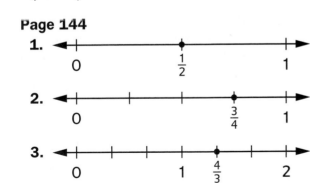

Page 165

1. Sample answer: The measurements are different because the hand of each person is a different width.

2. Sample answer: The measurements are different because the arm span of each person is a different length.

Page 175

1. 15 ft

2. 60 cm

3. 5 yd

4. Sample answer: 39 inches

$8\frac{1}{2}$ inches

11 inches

Page 179

1. 14 sq cm

2. 27 sq in.

3. 1 square meter, because 1 meter is longer than 1 yard (see page 173)

Page 181

1. Sample answer:

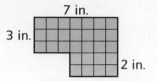

Area of blue rectangle + Area of red rectangle = 21 sq in. + 8 sq in. = 29 sq in. Yes, this is the same as the area found in the example.

2.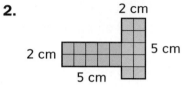

Area of blue rectangle + Area of red rectangle = 10 sq cm + 10 sq cm = 20 sq cm. The area of the figure is 20 sq cm.

3.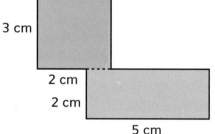

Area of blue rectangle + Area of red rectangle = 9 sq cm + 10 sq cm = 19 sq cm. The area of the shape is 19 sq cm.

Page 188

1. 12 hours; 12 hours; 12 hours; One time is A.M. and the other is P.M. Both have the same hour and minute. The elapsed time is 12 hours or half of a day.

2. 23 hours and 59 minutes

Page 190

1. **a.** Water and tomato juice

   b. 2 children

   c. 8 children

   d. 25 children

Page 194

1. 3 children

2. 3 children

Page 197

1. **a.** 15 children

   b. 1 child

   c. 3 children

Page 211

1. **a.** hexagon

   b. quadrilateral

   c. decagon

   d. octagon

   e. dodecagon

2. Sample answers:

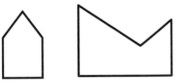

Page 217

1. One pair of parallel sides: pink trapezoid, orange trapezoid, and blue trapezoid

   Two pairs of parallel sides: green parallelogram (or green trapezoid)

2. Squares are also rectangles because they have all of the attributes of rectangles. Squares are parallelograms with four right angles.

Page 271

1. Bowling ball

2. Softball

3. 355 g

Page 283

649 words

Page 284

1. E, T, A, O, and N

2. J, Q, Z, X, and K

3. Answers vary.

Page 295

1. 787

2. 120

3. 330

4. 38

5. 5, 8, 11, 14, 17, 20

6. 10, 8, 6, 4, 2, 0

A

Abacus, 128, 130
Abbreviations
  in metric system, 166, 182, 183, 288
  for time, 184, 288
  in U.S. customary system, 171, 182, 288
Accuracy, 20–22
Addends, 108
Adding-a-group strategy, 47, 50
Addition
  basic facts, 108–109
  column addition, 118
  doubles, using, 110
  estimation in, 103, 116, 118, 256–257
  extended facts, 114–115
  grouping addends, 111
  helper facts, using, 110
  making combinations, 110–111, 115
  missing addends in, 255
  on number grids, 91
  partial-sums addition, 116–117
  practicing, 255–257, 261
  repeated addition, 26, 38
  strategies for, 110–111
  of whole numbers, 108–111, 114, 115,
    116–118, 261
Addition/subtraction facts table, 108, 109
Addition Top-It, 261
After noon (P.M.), 184. *See also*
        Before noon (A.M.)
Airline schedules (real-world data), 282
A.M., 184, 189
Analog clock, 186
Angles
  defined, 208
  measuring, 208
  parts of, 208
  in polygons, 212, 213
  in quadrilaterals, 212, 216
  right angles, 208
  in triangles, 214, 215
Animals
  clutches for (real-world data), 266–267
  growth of, 80–81
  migration routes for, 79
  predators and prey, 84
  survival of, 82–83
Arabic numerals, 127
Area
  defined, 176, 177
  finding, 176–181, 230–231
  of North America, 273
  precision in finding, 176
  of rectangles, 49, 178–179, 180–181
  of shapes, 176, 177, 180–181

skip counting to find, 178
in 2-dimensional shapes, 219
units for, 176, 288
using composite units to find, 178
using the doubling strategy to find, 49
Area and Perimeter Game, The, 230–231
Area models
  break-apart strategy, 51, 52, 60
  for doubling, 49
Arguments, 12–14, 34–35
Array Bingo, 232–233
Arrays
  adding-a-group strategy, 47, 50
  defined, 41
  division modeling with, 238–239
  for equal groups, 42
  for equal sharing, 43
  even numbers and, 71
  factor pairs and, 65–66
  multiplication modeling with, 25, 26,
    232–233
  odd numbers and, 71
  parts of, 41
  on Quick Look cards, 25, 26
  square numbers on, 71
  subtracting-a-group strategy, 48, 50
  turn-around rule for multiplication and, 25, 26,
    45, 66
Associative Property (of Multiplication), 47, 52
Attributes of polygons, 209, 212, 262

B

Back to Zero, 253–254
Ballpark estimate, 101
Bar codes, 88
Bar graphs, 191–192
Base-10 blocks
  modeling with
    expand-and-trade subtraction, 9–11
    partial-sums addition, 117
    representing numbers, 100
    trade-first subtraction, 121
  naming numbers with, 97
Base 10 system, 98, 126, 128
Base-20 system, 126
Base-60 system, 126
Baseball Multiplication, 234–236
Basic facts
  addition, 108–109
  division, 58
  multiplication, 57
  solving extended facts with, 114–115, 124
  subtraction, 108–109
Bead abacus, 128
Beat the Calculator, 237
Before noon (A.M.), 184

G

Index

Index